Praise for *The Sea Is R*

"The Sea Is Rising and So Are We is
once a primer for activists and an astute commentary
on a set of critical topics that even a seasoned climate
stalwart could benefit from. It takes on some really
tough questions—transformational change, how to
talk about the emergency, the need for a specifically
global politics of climate justice—and it does in a
manner that is both simple and sophisticated. It's
not an easy balance, but Kaufman pulls it off."
—Tom Athanasiou, author of *Dead Heat:*
Global Justice and Global Warming

"In *The Sea Is Rising and So Are We* Cynthia Kaufman
has provided us with a vital manual for confronting
the climate crisis and its root causes. Kaufman offers
compelling analysis, a comprehensive mapping
of the political landscape, and practical guidance
for action—all in a straightforward and accessible
manner. Most importantly, she offers hope."
—Tony Roshan Samara, program director of
land use and housing at Urban Habitat

"Cynthia Kaufman's *The Sea Is Rising and So Are We* is
a valuable overview of where we as a species are in the
existential fight to prevent catastrophic climate disruption.
It covers a lot, from the UN's Intergovernmental Panel
on Climate Change assessment of our situation to the
need for a personally supportive movement culture to
sustain our climate activism. It is an accessible, up-to-
date resource both for those who have been in the
climate fight for decades and those who know they
need to do so but haven't yet figured out how."
—Ted Glick, longtime climate organizer
and author of *Burglar for Peace*

"Cynthia Kaufman's *The Sea Is Rising and So Are We* challenges us to focus our attention on the powerful actors and structures that are at the root of our current climate crisis. In this moment of rapid transformation, Kaufman pushes us to see the reality of the situation we are in while providing concrete examples of actions that are already being taken and ways that people with diverse talents and interests can all contribute to creating a sustainable world. As a teacher of undergraduate courses on public policy, environmental politics, and community organizing, I welcome this work that seamlessly weaves together all of those elements in a profound yet accessible way."
—Lena Jones, board member of Center for Earth, Energy, and Democracy and political science instructor at Minneapolis Community and Technical College

"Racial justice and anti-capitalism need to be core to the movement to stop climate destruction. By focusing on challenging the entrenched interests that are driving human society toward destruction, this book points us toward the kinds of solutions that we need to throw our hearts and souls into with as much energy as we can mobilize."
—Eddie Yuen, coauthor of *Catastrophism: The Apocalyptic Politics of Collapse* and *Rebirth and Confronting Capitalism: Dispatches from a Global Movement*

The Sea Is Rising and So Are We

A Climate Justice Handbook

Cynthia Kaufman

BTL

PM

The Sea Is Rising and So Are We: A Climate Justice Handbook
© 2021 Cynthia Kaufman
This edition © 2021 PM Press

ISBN: 978-1-62963-865-2
Library of Congress Control Number: 2020947292

First published in Canada in 2021 by Between the Lines
401 Richmond Street West, Studio 281, Toronto, Ontario, M5V 3A8,
Canada
1-800-718-7201
www.btlbooks.com

Canadian Cataloguing in Publication information available from
Library and Archives Canada

Cover by John Yates / www.stealworks.com
Interior design by briandesign

10 9 8 7 6 5 4 3 2 1

PM Press
PO Box 23912
Oakland, CA 94623
www.pmpress.org

Printed in the USA.

This book is dedicated to all the brave
and passionate people who are working
to build a just and sustainable future.

Contents

Foreword

Bill McKibben

This book is wonderful for its straightforwardness and simplicity. Cynthia Kaufman doesn't claim, I think, to break new ground, but the truth is that new ground doesn't need to be broken. We know most of what we need to know about climate change—its causes, dangers, and solutions. What we need is a commitment to fight and a plan for making that fight effective. What we need is a way past the barriers—structural, political, psychological—that keep us locked on what is frankly a suicidal path. And these are precisely what this book supplies.

I wrote what is often called the first book on global warming, way back in 1989. Even then we knew pretty much what would happen if we kept burning gas, oil, and coal. But we also believed, I think, that if the facts were known, then the powers that be would begin to act with the speed and courage the science required. That turned out to be wrong. Though we quickly won the scientific argument, we continued to lose the fight, because the fight wasn't about data and evidence. Instead, like most fights, it was about money and power, and the fossil fuel industry possessed those in quantities sufficient to carry the day.

Their power cost us thirty crucial years. But by now we see the scope of the crisis so clearly (see it in the flames, the floods, the rising oceans) that huge numbers of people want to act. And increasingly we see the villains in this drama for what they are: we understand that the oil companies, and the giant banks that lend them their money, are willing to break the planet in order to extend their business models a few more decades.

That clarity doesn't make the task at hand *easy*, but it does make it simple. And Kaufman's genius lies in explaining how to go about that task. How to join together in the multiracial, intergenerational coalitions that can hold power accountable; how to bring the message of a positive future that can motivate more people to act; how to break through our own psychological obstacles and free ourselves to act as we must.

I don't know if we will succeed—the scariest element here is the short time that physics is giving us to act. But I do know, from years of organizing around the world, that we will fight. And I know that books like this one—based on real-world experience of many battles—will help us understand what that fight must look like. It is a gift to all of us at work in this battle (and a great gift to give to anyone you want to bring on board).

Acknowledgments

I want to thank everyone working for a just and sustainable future. Special thanks go to those who read early drafts and gave advice for the book: Anna Goldstein, Carlos Davidson, Amy Merrill, Tom Athanasiou, Jacques L'aventure, Anna Goldstein, Patrick Reinsborough, Brian Malone, Bill McKibben, Marian Mabel, Dan Fuchs, Leonard Skylar, Zoe Volpe, Keris Dahlkamp, and Jamie Henn.

Introduction

"Hope is belief in the plausibility of the possible as opposed to the necessity of the probable."
—Moses Maimonides

Near where I live, in Pacifica, California, there is a big piece of land right at the ocean that was once slated to become a giant freeway interchange. That plan was stopped by local people who saw no need for increased freeway capacity.

When I moved to town fourteen years ago, that land was covered in muddy, disconnected dirt bike trails, almost inaccessible to walkers, and overgrown with invasive weeds. A few years after that, the land was acquired by the federal government, replanted with native plants, and developed with walking and biking trails. It is now a thriving home for countless bugs and butterflies that rely on the native plant species to survive, and it is becoming a better home for the endangered San Francisco garter snake, one of the most beautiful snakes in the world. From the blufftops of Morey Point you can see lines of Brown Pelicans, those iconic birds that were taken to the brink of extinction by DDT and brought back by regulation inspired by Rachel Carson's *Silent Spring*.

The park is also a place of refuge for a racially and economically diverse group of people who come from far and wide to walk, get a break from the city, exercise, and enjoy spectacular views of the ocean. The park's open spaces are free and belong to everyone. The park uses federal tax money to hire a very diverse set of employees and helps paid interns on their paths to meaningful careers.

One of my favorite things in life is watching a devastated landscape be turned into something thriving, beautiful, and socially sustainable. It helps heal me from the sense of hopelessness I often feel in the face of the climate crisis. I take heart in Arundhati Roy's statement, "Another world is not only possible; she is on her way. On a quiet day, I can hear her breathing."[1] For that new world to really be brought into existence, it needs especially, to be able to breathe.

What will it take to heal that seven-mile-thick strip of air that provides the foundation for life on earth, which we call the atmosphere? We are living in a time of unprecedented change, much of it for the worse. The possibility is real that human society will continue on the path of destruction, and the world will become a permanent war zone, as our land becomes inhospitable for farming, we are subject to more disasters, social systems become unstable, and people fight for survival.

Also real is the possibility that we will act quickly enough to heal the atmosphere that keeps all of our ecological systems functioning, and in the process we may actually make our societies more livable and socially just. And, of course, we are most likely to end up with some of both. There has already been irreparable harm, and there is more destruction certain to come. How much more we don't yet know. There also may be some very significant positive changes in society that come as a result of dealing with the climate crisis. Which future we end up with is an open question, and it is one that we all have a part in determining, with the acts we take in the next short period of time.

Many of us are in a deep state of despair as we face the reality of how far along this crisis is and how entrenched the political forces are that keep us from making this transition as quickly as we need to. The climate is already hitting some of the tipping points we have been warned about. For example, as the polar ice melts, the ocean becomes blacker and absorbs more heat, rather than reflecting it. Similarly, as draught leads to fire, those fires emit more greenhouse gases. The situation we face is urgent, and we need to reduce emissions dramatically and immediately. That is a tall order when we realize that global emissions are continuing to rise. We all find our own way to deal with these realities on an emotional level. For myself I have found that it is important to simultaneously be open to knowing how bad the hand is that we have been dealt while also focusing my attention on those places where I can make a difference.

There are different roles for different people in moving to a society that can sustain a healthy atmosphere. There are engineers who are inventing better heating and cooling systems, better electric buses, and better appliances. There are people in businesses working to save money by using less energy. There are artists who are helping us understand what we are up against and to connect us with our passions to build a better world. There are investors putting money into green energy and sustainable practices. There are politicians working on regulations mandating stronger vehicle emissions standards. There are individuals who are changing how they live to lower their personal carbon footprints and encouraging others to do so as well.

All of those things are crucially important, and if you are doing them you should feel proud that you are helping to bring the transition to a sustainable society. But there is one missing element if those are the only things that are done: challenging power. There are entrenched powers in society keeping things from changing as quickly as they need to change. As I write this, fossil fuel companies are still building new infrastructure, spreading misinformation, and trying to protect their ability to sell the resources they count as assets. All around the world, governments are subsidizing fossil fuel production to the tune of around over $4 trillion a year, with the US contributing $649 billion a year.[2] That is more than the US, the largest military power in the world, spends on the military.

Politicians in many countries still take campaign money from fossil fuel companies. And in the US, fossil fuel titans, such as the Koch brothers, have poured billions of dollars into partisan gerrymandering, and other schemes to make it difficult for candidates to get elected who want to take bold action for the climate. Those billionaires, and the major fossil fuel companies, have used that money to spread misinformation around the climate crisis. That work has lulled millions of people in the US into a sense of complacency, which slowed down a political response by several decades. It has also helped make our political system unable to take bold action.[3]

In order to heal the atmosphere, which all living things depend on, and which holds together all of the systems that make life on earth possible, we need to disrupt the social and political systems that are getting in the way of a rapid transition to a healthy and just society. The climate justice movement is focused on challenging the powers that keep us on a path of destruction. It focuses its attention on what is needed to have a just transition, that is, a transition to a society that is both environmentally sustainable and socially just.[4]

Because we live in a world of multiple and intersecting forms of power, that work requires careful attention to capitalism, racism, and sexism. The climate crisis is a threat multiplier. Those with more money can move up hill as the coasts become unlivable. And the continuing racial discrimination in lending will mean people of color will be less likely to get the loans needed to move. As hurricanes hit Puerto Rico with increased intensity, its status as a colony will mean it is less able to rebuild. As large parts of Africa become unfarmable, the women who do not hold title to land will be harder hit than the men who own land and who are able to leave the land and move to cities. The 2017 and 2018 fires in California were devastating for people from all socioeconomic classes. But wealthier people had homeowner's insurance and most have been able to rebuild. Renters, and those owners without adequate insurance, have been left homeless or have had to leave their communities.

People with resources, social status, and political power are more protected from the devastating impacts of the climate crisis than are those with less of these things. The needs of frontline communities, those who work in fossil fuel extraction and don't have other job opportunities, those who live in communities devastated by the drilling, mining, and refining of fossil fuels, and those who are exposed to the impacts of disasters without insurance to cover their losses, need to be front and center of our analysis of what we are facing and the best choices for policies.

Much of the work needed to remove the barriers to the adoption of the practices needed to switch from extractive and exploitative economic and political systems to ones that meet human needs requires that we challenge entrenched powers. We need forms of action that work hard for goals that run counter to the system that is designed to keep moving along in its preprogrammed ways.

Activism is not always disruptive, but its goal is to challenge the inertia behind a world on a path of destruction. People engaged in activism can bring a resolution to a city council to ask the city to build affordable housing near transit lines. And the city may agree to that without a fight. But activists must also be prepared to bring pressure to bear when that is what is needed. Maybe they will need to mobilize to bring more people to the council meeting, maybe they will need to talk individually to council members. Maybe they will need to work to elect different members who are more willing to act to protect the climate. In all of these cases, those engaged in activism have goals they want to achieve, and they don't just allow the system to carry on as if the climate crisis didn't matter, and as if the needs of everyone were automatically in the equation for a transition to a sustainable society. Instead, they mobilize grassroots power to achieve their goals.

While this book will be helpful to anyone wanting to understand the landscape of climate action, it is specifically directed to those willing to strategize about how to have the most impact, willing to engage in actions that no one told them they were authorized to engage in, and willing and ready to challenge the power in existing systems. They are the people who are prepared to disrupt business as usual to get a different outcome, to break the inertial pull of our entrenched systems of power.

For many years, those of us working to transform society away from a dependence on fossil fuels were clear what we were against, but not entirely clear what we were for. Twenty years ago, when I first started thinking about these questions, I wasn't sure if we could live sustainably in cities, if it was possible to have a sustainable society with a large population, or if we could still have lives of creature comforts like heating and cooling, in a sustainable world. And we weren't sure how to go about transforming all of society's systems to get to sustainability. Now all these years later, those question have been answered, and the answers are almost entirely positive.

Right now, the technology is available to feed the world at the population levels were likely to see as the global population is on track to stabilize in the coming decades. The technology is available for us to live urban, cosmopolitan lives of comfort and pleasure, while not emitting greenhouse gases. And a vision is developing of a sustainable world which has much more joy and human pleasure than are possible in the current systems based on inequality, racism, sexism, heteronormativity, consumerism, and corporations out of control.

Those of us close to the world of climate action know that huge changes are already happening, as cities develop sustainable and egalitarian systems of transportation, as countries invest in renewable energy, and as regenerative agriculture is developing. But we also know that those better systems won't naturally "outcompete" the fossil fuel–based ones for as long as our political systems remain captured by the forces of free market capitalism. As we will explore further in chapter 2, capitalism is a social arrangement that allows major social decisions to be made by for-profit businesses, those business operate through markets which are shaped by those with power, and it allows "externalities," such as fossil fuel companies being able to use our atmosphere as a dumping ground for greenhouse gases.

We know that to get the changes needed, at the speed and scale we need, governments will have to be captured by those interested in a just transition to a sustainable society which serves human needs. And while that may seem wildly unimaginable, especially in the US, where our government is controlled by the interests of the 1 percent, something like this has happened before and it needs to happen again.

Much of the excitement behind the concept of the Green New Deal that was proposed in 2019 was that it reminds us of a time in the not-so-distant past when social movements were so strong that a reluctant President Roosevelt enacted a huge series of reforms that completely changed the face of the US. As a result of those movements threatening social disruption, the government stepped up and developed social security, occupational health and safety standards, a requirement for a weekend, health care for the elderly, and much more.

The New Deal included significant compromises with Southern Democrats who at the time would only go along if farm and domestic labor were left out of that New Deal, and if states would be allowed to administer it in ways they saw fit. These compromises allowed for racist implementation of many of the programs of the New Deal, such as in lending for housing and the exclusion of farm and domestic labor from its protections.[5] When Alexandria Ocasio-Cortez and Edward Markey proposed their fourteen-page manifesto in 2019 calling for a Green New Deal, they were well aware of those failings. But they also saw the reference to the New Deal as a reminder of how much can happen when society is mobilized to work for rapid and deep social transformation.

Their Green New Deal plan calls for the federal government to act as a vector of massive social transformation in a short period of time. And their vision is being filled out with a number of more detailed but equally ambitious specific plans. In 1997 most countries of the world signed the Kyoto Protocol, the first major international agreement to reduce greenhouse gas emissions. That plan was an important first step, but the people who were paying attention to the science knew that the agreement was not ambitious enough to actually solve the problem. None of the international agreements signed since that time have been adequate, meaning that they do not ask countries to agree to commitments that are enough to achieve a sustainable world. The Green New Deal proposal was the first one to get attention in the US to call for actions that are ambitious enough to actually stabilize the climate. The plan also calls for a just transition that eliminates greenhouse gas emissions while also addressing other interrelated social problems.

Plans for a just transition call for investment in communities that need good union jobs. They expand the power of working people to advocate for their need for good jobs. They ask that policymakers listen to the voices of people who have lived at the front lines of racial exploitation and environmental injustice for decades, that they center the interests of indigenous people. They call for restoration of communities devastated by underinvestment and toxic pollution.

Calls for a just transition also need to take seriously the fact that US, along with the other wealthy nations of the world, has emitted most of what is already in the atmosphere and has disproportionate power over the transnational institutions, such as the United Nations, the World Bank, and International Monetary Fund, which have tremendous influence over the world's flows of capital. The US needs to work with other nations to invest what is necessary for the countries of the Global South to also have the resources necessary to make the transition.

The original New Deal came about as a result of massive social upheaval growing out of the Depression, with people demanding jobs, housing, and social safety nets. The work of those activists translated into huge government programs to meet social needs. And, while the New Deal included many programs that excluded people of color from protections and opportunities, there were many projects that hired people of all races to build beautiful and useful projects that have lasted for decades.

About one hundred miles south of where I live is Pinnacles National Park. It is a set of dramatic jagged red-rock peaks. Flying over the peaks are California Condors, one of the largest bird species in the world, brought back from the brink of extinction, like the Brown Pelican. The trail along the top of those peaks is so jagged that a set of steps and railings were built during the time of the New Deal to make them passable. The park also includes a small reservoir that is surrounded by the beautiful stonework that is the hallmark of New Deal landscaping.

Whenever I come across that beautiful stonework in public spaces all over the US built during the New Deal, I think of the legacy left by that investment in our shared future. It gives me that sought-after feeling that another world is on her way. People have come together before to build a common future, and we can do it again.

We are at a time of renewed upheaval, as millions of people are realizing the urgency of the task before us to build a just transition to a sustainable society. Many are becoming involved in action to solve the problem. Yet, unfortunately, addressing the climate crisis brings us into dialogue with a whole range of technical and scientific issues, and it involves engaging with complicated social processes that need to be changed. All of that can be intimidating and can slow people's approach to involvement in the movement.

This book is intended as a ramp up to involvement with the movement for a just transition to a sustainable society. It also offers information and analysis that should be useful for more seasoned climate activists. It attempts to explain the things that one comes to hear about as one becomes involved with climate action. Taking climate action does not require expertise, but we can be more effective and more confident when we understand the context of our work. And we can be more effective, and our climate action can be a more positive part of our lives, when we understand how the movement works and what the best places in it are for each of us.

Chapter 1 is an overview of the forces we are up again, both scientific and political. Chapter 2 lays out a vision of a better world that we can build as we transition from a society built on burning fossil fuels and on exploitative relationships, to a society based on meeting human and ecological needs. Chapter 3 focuses on the idea that how we talk about what we are doing as we face the climate crisis will have big impact on the willingness of others to join us. Chapter 4 is an overview of the large-scale social transformations that need to happen, and that are happening, to build a just and sustainable world. Chapter 5 is intended to help the reader understand how social change happens. Chapter 6 is an introduction to the main policy tools that people working for action are trying to implement. Chapter 7 explains the ways that the movement for a just transition is organized, to help the reader find the flavor of climate action that best suits their interests and personality. Chapter 8 ends the book with a look at how we can maintain our mental health and happiness in the face of most urgent and large-scale crisis our species has ever faced.

I hope this book will help you find your place in engaging in effective action to build a sustainable and just society and will help you invite others to find their places in the work. And I hope it will help you find joy in that work.

CHAPTER 1

What We Are Up Against: Science and Politics

Getting the world off fossil fuels and restoring the health of our atmosphere requires massive transformations in how society functions. The bad news is that it needs to be done in a matter of years, not lifetimes; that much damage has already been done; much of that damage, such as the extinction of species, is irreversible; and the political systems of most countries in the world, especially in the US, are not organized in ways that make those transformations easy.

The good news is that all around the world the transition is already happening, and the things we need to do to solve this problem will help us build a society that is better in virtually every way. The transition will require unprecedented effort, and will lead to serious changes in how we live. But because there are so many good things that come from doing that work, and so many ways that what we are aiming for is a better world, we should not see the transition in terms of sacrifice.

Rather, addressing the climate crisis requires that we make our societies more democratic and inclusive; that we develop our cities in ways that will make them more joyous places to live; that we farm in ways that are better for farmworkers and farm families; that we eat in ways that are healthier for us; that we move away from the use of internal combustion engines, which cause heart and lung disease, especially in low-income communities of color; and that we change our political systems such that they are not run in the interests of corporate profits.

In order to keep the earth habitable for our species and others, we need an extremely rapid reconfiguration of society and how it functions. The scale and speed we need cannot be overestimated. Yet the goal we are headed for is almost entirely a positive one.

The United Nations Intergovernmental Panel on Climate Change (IPCC) 2018 report set 2030 as a benchmark by which time the world needs to reduce emissions to 45 percent below 2010 levels, if we hope to achieve its longer-term goal: to completely decarbonize the world economies by 2050. And that goal is necessary if we are to keep the world to a 1.5°C level of warming, which scientists say is the absolute maximum possible to have a sustainable atmosphere. There is an almost complete scientific consensus on those claims, and to the extent that they are criticized by scientists, it is on the grounds that they are not ambitious enough. Those 2030 and 2050 benchmarks have helped focus the attention of many people around the world on clear and specific goals. Achieving those goals will require millions of us to be involved in activism to get our political systems to change course. In the US and in many other industrialized countries, the political system has been dominated by fossil fuel corporations, and has been operating more generally in the interest of wasteful and destructive capitalism for a few hundred years.

The task scientists have put before us is, as quickly as possible, to stop burning fossil fuels, stop emitting other greenhouse gases, stop destroying forests, and to start planting trees and doing other things to pull carbon out of the atmosphere. The science of why and how quickly we need to do that and what the consequences are if we don't are incredibly clear are based on strong scientific consensus.[1]

What Is Happening with the Atmosphere?
Since the beginning of the industrial revolution, human beings have been putting CO_2 (carbon dioxide), methane, refrigerants, and other gases into the atmosphere at a rapid

rate. Those gases are building up and trapping the sun's heat on earth. The atmosphere is like a thick blanket and the extra gases make the blanket thicker. Those CO_2 molecules can stay in the atmosphere for thousands of years.[2] Thirty years ago, most people called the crisis we face "global warming." But activists and policymakers realized that this gave a false impression of what we are facing. If all the greenhouse gases did was to warm the earth, it might be good for agriculture in some parts of the world, like Canada. The reason we now call it catastrophic climate change, and not just global warming, is that a warmer atmosphere works differently than the atmosphere does without those extra greenhouse gases in it.

Terrible Things That Happen as a Result

Scientists have been warning for years of extreme weather events, drought followed by extreme rain, and the disruption of weather patterns. So much greenhouse gases have been put into the atmosphere at this point that there is no way that we can avoid having some pretty terrible consequences. We have already seen devastating impacts from the warming of our atmosphere. All around the world droughts are causing wildfires and destroying the ability of farmers to grow their crops. The polar ice sheets are melting. And we are experiencing devastating hurricanes.

No matter what we do at this point, the glaciers will continue to melt, and the seas will continue to rise. The weather will be more extreme and unpredictable for a good long time to come. Many countries, especially in Africa, will have their agriculture disrupted by drought. What we are fighting over is how bad it will be and what kind of society we have on the other side.

There is also worry that as the problem gets worse, tipping points will be reached that will make bad things worse. For example, as the arctic sea ice melts, this exposes more open ocean which is darker, less reflective, and more absorptive of heat than ice. Similarly, as the permafrost in the northern latitudes melt, it gives off methane, again, making the problem worse.

The World Health Organization estimates that between 2030 and 2050, 250,000 people will die per year from climate change related causes.[3] And the climate crisis is one of the major factors leading to global migration worldwide, rivaling the impacts of war. And in many places which have recently experienced war, such as Sudan and Syria, climate change has been a partial cause, as people are unable to thrive under changing circumstances, such as drought and flooding. The National Network for Immigrant and Refugee Rights argues that there are already "more than 25 million people around the world displaced due to climate change."[4] Those increased levels of migration then are related to political crises as nativist politicians attempt to close borders to those seeking refuge from places that are becoming difficult to farm, such as rural Guatemala and Honduras.

The International Consensus on the Goals

The Intergovernmental Panel on Climate Change (IPCC) is a body of scientists formed in 1988 to track the latest scientific consensus on climate change. That body has wanted to ensure that its statements were never criticized for being exaggerations, in a world full of misinformation and denial. As a result, it has been criticized by some as overly cautious and conservative in its statements. The IPCC regularly puts out reports which are important drivers of the climate negotiations that take place at international meetings to set policy. The IPCC has argued that 350 parts per million (ppm) CO_2 in the atmosphere is the safe level. As of this writing, in 2019, we are at 415. To keep the planet to a 2°C level we need to stabilize the atmosphere at 450 ppm CO_2.[5]

An international meeting in Rio De Janeiro in 1992 established the United Nations Framework Convention on Climate Change (UNFCC). That treaty continues to evolve at regular international meetings. At the UNFCC meeting in 2015 in

Paris, almost all of the countries of the world agreed to the goal of holding the rise to 2°C and to staying as close to 1.5°C as possible. In 2019 the IPCC put out its dramatic and clear 2030 and 2050 goals to achieve those levels of warming: a 45 percent reduction in emissions by 2030 and complete decarbonization by 2050.

The Paris Accord was the first time that the countries of the world agreed to something that was close to the levels of emissions reductions that were needed to keep the world livable, and to stabilize the climate. That agreement did not include binding regulations and mechanisms to sanction countries that did not keep to their commitments. What it does include is a ratchet, whereby countries are encouraged to increase the ambitions of their goals over time. The Paris Accord opens countries to pressure from activists to make reductions that hit those targets. If we can hold the rise to 1.5°C, the earth is likely to remain habitable by large numbers of human beings.

In her book *Climate Justice*, Mary Robinson, former president of Ireland and former UN High Commissioner for Human Rights, argues for the need to hit the 1.5°C goal:

> More than 1.5°C above 1880 levels would lead to the loss of 90 percent or more of all coral reefs. An increase of 2°C would almost double current water shortages around the world and lead to a massive drop in wheat and maize harvests. The vicious heat waves that we experience today would become the norm, and the inundation of coastal cities like that of Houston, Texas, in August 2017 would become routine, forcing millions of people to lose their homes.[6]

Reaching the 1.5°C goal will require huge investments in renewable energy and planting trees. Most of those investments need to come from the wealthy countries of the world, which have emitted 79 percent of the greenhouse gases which are already in the atmosphere.[7]

That 79 percent of emissions in the atmosphere was not just put there by the actions of all of the people in the wealthy countries of the world. Fully one-third of total emissions were emitted by just twenty companies. Richard Heeded of the Climate Accountability Institute created a list of the top greenhouse gas emitters since 1965, when emissions began to rise significantly.

> The top 20 companies on the list have contributed to 35% of all energy-related carbon dioxide and methane worldwide, totaling 480bn tons of carbon dioxide equivalent ($GtCO_2e$) since 1965. Those identified range from investor-owned firms—household names such as Chevron, Exxon, BP and Shell—to state-owned companies including Saudi Aramco and Gazprom. Chevron topped the list of the eight investor-owned corporations, followed closely by Exxon, BP and Shell. Together these four global businesses are behind more than 10% of the world's carbon emissions since 1965.[8]

While people used the gas that those companies sold, it is important to remember that a key driver of the climate crisis, and of the slowness of the transition away from burning fossil fuels, has been the political work these companies have done to slow the transition away from fossil fuels. Many of them have spent billions of dollars trying to undermine belief in the scientific consensus, and they have invested in politicians who work to make policy according to their interests.[9]

For a few years, activists talked about allocating fairly the rest of the world carbon budget, such that the poor countries of the world would be able to emit more as they moved their populations out of poverty. Now it is clear that the world has no carbon budget. We have already emitted more than is safe. Now the question is how countries can get to net zero, meaning that for the greenhouse gases they emit, they also pull them out of the air, by planting trees and doing other things to sequester carbon, to have a total of zero emissions.[10]

The organization EcoEquity, in its analysis of the national pledges of climate action, argues that, if it is to do its fair share in the shared global effort of stabilizing the earth's climate system, the US should, by 2030, cut its emissions by far more than 50 percent, and its 2050 goal needs to be well below net zero. At the same time, it will need to contribute hundreds of billions of dollars to the UN's Green Climate Fund, to help poor and developing countries do the same. By 2050, if humanity is to hold the warming to 1.5°C, net global emissions will have to be almost zero, and that means that all countries, rich and poor, are going to have to cooperate to draw down their emissions at an extremely challenging rate.

Mary Robinson argues that the transition to renewable energy will have huge positive impacts on many countries in the Global South.

> Nearly three billion people still live without access to clean cooking. Instead, to cook they rely on high-polluting solid fuels—wood charcoal, animal dung, and crop waste—with fumes that kill more than four million people every year. . . . Providing electricity to the 1.3 billion people who lack access across the developing world remains one of the largest challenges on earth. . . . Once villages are electrified and have access to clean cooking, they will have access to better health care and schools with electric light where children can study longer.[11]

Many of the countries of the Global South are committed to using clean energy as they move people out of poverty. Robinson writes of Ethiopia, that "although two thirds of its people currently have no access to electricity, Ethiopia has pledged to achieve ambitious emissions reductions and invest in renewable energy by 2025."[12]

Costs of Doing It Versus the Costs of Not Doing It

In conversations around facing the climate crisis many people balk at the price tags put on the work that needs to be done to invest in the changes that need to be made to build the new sustainable infrastructure we need all around the world, and also to build up resiliency to the problems already expected to happen as the seas rise and the weather becomes more extreme.

In 2006, British economist Nicholas Stern published an influential analysis, *The Economics of Climate Change*, in which he showed that, leaving aside the other terrible consequences of increased greenhouse gases in the atmosphere, the economic costs of making a transition to a sustainable society were a small fraction of the cost of dealing with the crises that would follow from weak action.[13] Cristina Figueres, a Costa Rican diplomat and ex-head of the UNFCC put that choice starkly: "We will move to a low-carbon world because nature will force us, or because policy will guide us. If we wait until nature forces us, the costs will be astronomical."[14]

We have already seen in the US that Superstorm Sandy, which devastated much of New York City, cost $65 billion. The combined cost of climate-related disasters worldwide from 2017 to 2019 was $650 billion.[15] And those costs are just the price of repairing the damage. They don't calculate the trauma of those impacted, the devastation of people's sense of safety, the loss of life, or the devastating political impacts of instability.

Whether we engage in strong swift action, or allow business as usual to unfold, or do something in between, the climate crisis will cost a tremendous amount. Forty-four percent of the world's population lives within 150 km of the sea.[16] With the current levels of emissions, those coastal cities will need to be reengineered in significant ways. The subway systems of major cities will need to have investments of billions of dollars. In the San Francisco Bay Area, all three major international airports will be underwater with predicted sea level rise.

The more the seas rise, and the more we have hurricanes, fires, and droughts, the more expensive it will be to recover from these crises. It isn't a question of whether we can afford to do the things needed to make the transition from fossil fuels. Asking the narrowest financial question: which path is the least expensive, the answer is obvious: swift, adequate action will cost less. And the basic moral question, of which approach will save more lives, lead to less forced migration, lead to less species loss, and lead to a better future, the choice becomes even more obvious.

If It Is Better and Cheaper to Transition to a Sustainable Economy, Why Hasn't It Happened Faster?

The problem we face is not one of total costs versus total benefits. The problem is who stands to lose, who stands to gain, and who controls our political system. The world's major fossil fuel companies have trillions of dollars' worth of assets that they count on their books and which help determine the value of their companies. It is estimated that 75 percent of the raw resources that these companies count as assets will need to go unburned to stay within a 1.5°C warming. Once it becomes clear that those assets will never be able to be burned, those companies stand to lose almost everything.[17]

In fact, the organization Asset Retirement Obligation Watch has argued that in the US, many states and the federal government have laws on the books which require energy companies to be responsible for retiring their old wells. If we were to subtract from these companies the value of the assets they should not be able to burn if we are to remain near a 1.5°C goal, and we further subtract the money they should be required to pay for closing wells and cleaning up after them, those companies are all already bankrupt.[18]

So of course those companies are fighting to the death to keep selling and having us burning their assets, and they are fighting to avoid liability for cleanup. And in the US, they can still count on the support of the majority of Republican legislators, and many Democrats as well, to do their bidding. This disconnect between what is good for society and what is good for a particular power holder shows up at every level of society.

The Union of Concerned Scientists has estimated that there is $1 trillion worth of property in the coastal zones in the US that is likely to be impacted by sea level rise. That constitutes a real estate bubble about to burst, as fewer buyers become interested in buying those zones.[19] People have begun to talk about climate gentrification, where land on higher ground that used to be less desirable and has often been home to low-income communities of color is becoming more attractive to wealthier buyers. As the sea rises, much land right at the coast will need to be abandoned, or huge sums will need to be invested to protect it.

In my little town of Pacifica, we had a progressive city council that wrote a realistic plan for how to deal with the rising sea. The local, state, and national real estate organizations put over $100,000 into one of our local elections to defeat that council. They were very aware that taking a realistic look at what was coming with a rising sea would have an adverse impact on the value of property in the coastal zone. The real estate industry would like to keep making as much profit as possible for as long as possible, so they paid to elect a council that would defer that conversation. In 2012 North Carolina passed a law against planning for sea level rise.[20]

The concept of climate justice asks us to always keep issues of power and equity in mind when analyzing what is going on. Nicholas Stern was right that a transition to a sustainable economy is much cheaper than dealing with the consequences of "business as usual." But the fact that we aren't making that transition as fast as needed is not a sign that people are stupid. Rather, it is a result of the ways that our political systems are controlled by those who profit from an economic system based on fossil fuels, consumerism, and large-scale agribusiness.

The forces that continue to profit from each of those aspects of our economies are each a major brake on a just transition to a sustainable economy. And no amount of asking them to get with the program will change that situation. Rather, they need to have their power over social decision-making processes taken away. And that is where climate justice activism is most important.

The Political Realities We Are Up Against

The political work we need to do to achieve the transformation to a just and sustainable society takes place in a broken political system which is embedded in a devastated public sphere, where people are, for a variety of reasons, deeply cynical and mistrusting, where a sense of fear and hatred permeates much political discourse. Much of the Global North is being polarized into two old and unsustainable political camps, while many of us are fighting to open up a third possibility.

On the one hand is the dominant liberal consensus which believes in science, but which also believes in unlimited capitalist expansion and a "free market"-based global economy. People in that camp want business as usual, and when it is interested in sustainability, it wants things to be largely the same but without fossil fuels. In the US that view is represented by the mainstream of the Democratic Party. It is wedded to economic growth, progress, and a rationalist worldview, which sees nature as something to be used for production. These procorporate Democrats and their supporters have believed that a deregulated global economy could bring prosperity for all. They tend to see themselves as forward-looking in comparison with backward-looking ethnonationalists.

Many of these procorporate elites have supported getting the world off fossil fuels, but they want to leave the wasteful consumerist economy in place, and they show little concern for the disparate impacts of that economy on poor people and people of color. The economic system that they have supported includes free trade deals that favor corporate interest and that don't take seriously the ways that the procorporate globalized economy has been devastating to communities in the Global North as well as in the Global South. They have not worked to protect rural and rustbelt communities from the devastations of corporate-led economic globalization. And so that procorporate political elite, has lost much of its credibility and political power to the ethnonationalists.

On the other side of the existing political equation, is an emerging toxic ethnonationalism, which looks backward to an ethnically unified nation, and which has a fantasy of national control, ethnic homogeneity, and cheap plentiful fossil fuels to run a consumer economy. The forces funding that backward approach to politics are generally interested in a deregulated economy which allows for as much profit-making as possible, and it convinces people that the forces looking to regulate fossil fuels are the globalist forces wanting to undermine their sovereignty and sense of a national home. In the US, this is the politics of the Republican Party and its base in the one-third of the population who live in fear of being replaced by a racialized other and who have come unmoored from discourses based on fact. These people have tended to vote with the political party that also wants to keep us hooked on fossil fuels for as long as possible.

Getting the US off fossil fuels and on a path to sustainability requires a realignment of politics away from these two options, toward a set of goals that will offer all of us a sense of home and belonging, which will develop a sense of trust in shared institutions, and which will position people in the US as members of a global community, fighting for well-lived lives for everyone in all parts of the earth that is our home.

In his book *Down to Earth* French philosopher Bruno Latour helps us understand the nature of the political moment we are in and offers a map toward a political realignment that will make it possible to achieve a sustainable world.[21] Latour puts the climate crisis, the global migration crisis, and the transnational growth of right-wing nationalism into one clear framework. He argues that the rise in right-wing nationalism all around the world is related to the sense of insecurity people are experiencing in many countries. The past fifty years has seen a consensus among global leaders that more trade and less spending on social goods, like schools and health care, would lead to prosperity for all.

This view, shared by the dominant forces in the US Democratic and Republican parties and by global elites all around the world, is often referred to as "neoliberalism." "Neoliberalism" refers to the idea of "free trade" or trade liberalization. It is a confusing terminology, since conservatives are often more neoliberal than liberals. The "neoliberal consensus" has led to a tragic undermining of local systems of agriculture in much of the Global South, and regimes that allowed business to run roughshod over the needs of citizens in all of the countries where it has prevailed.[22] This has led to a period of deep insecurity. When the climate crisis is added to that, we see that we are up against a global crisis where many people turn to ethnonationalism as they search for something to ground their sense of self in, and also as a response to those forces of global capitalism which are destroying their ways of life.

Latour writes,

> We can understand nothing about the politics of the last 50 years if we do not put the question of climate change and its denial front and center. . . . Migrations, explosions of inequality, and New Climatic Regime: *these are one and the same threat.* Most of our fellow citizens underestimate or deny what is happening to the earth, but they understand perfectly well that the question of migrants put their dreams of a secure identity in danger.[23]

Latour sees the political question of our time as how to "reassure and shelter all those persons who are obliged to take to the road, even while turning them away from the false protection of identities and rigid borders."[24] By this he means people who have lost a sense of place and belonging and the security of a sense of home, as well as those who are literally on the road.

Latour is a philosopher of science and is interested in the ways that facts come to be created through by social practices. He isn't a postmodern relativist who argues that all knowledge is illusion. Rather he is a considered pragmatist who asks what the conditions are under which we can create forms of knowledge that work for us. He argues that we are living in a time of epistemological crisis, a crisis in the production of knowledge, whereby whole segments of our populations have given up on the idea of truth grounded in facts. But rather than tearing his hair out about how stupid people are, he digs into why this is happening. "Facts remain robust only when they are supported by a common culture, by institutions that can be trusted, by a more or less decent public life, by more or less reliable media."[25]

Latour argues that mainstream liberal globalist ideas about the modern global economy have contributed to this epistemological crisis:

> Before accusing ordinary people of attaching no value to the facts of which so-called rational people want to convince them, let us recall that, if they have lost all common sense, it is because they have been masterfully betrayed. To restore a positive meaning to the words "realistic," "objective," "efficient," or "rational," we have to turn them away from the Global, where they have so clearly failed, and toward the Terrestrial.[26]

For Latour, the Terrestrial is our home, it is the place where we are wrapped up in a variety of crucial life and death relationships, such as the health of the atmosphere. Latour hopes that a focus on the Terrestrial will help us see the deep linkages between social and ecological struggles, which will lead to an increased political commitment to restoring the matrix in which human beings live our lives.

At the end of the book Latour writes as a European aware of Europe's responsibility, over the past few centuries, for the devastation of the homes of billions of people in the Global South. He calls on Europeans to take responsibility for their locale, as an "experiment in what it means to inhabit an earth after modernization, with those who modernization has definitely displaced."[27] He then calls on all of us to take seriously the healing of the Terrestrial zones we inhabit.

In many parts of the world people are "rewilding," doing things like reintroducing top predators and allowing the natural environment to reemerge in a healthy form. Millions of people all around the world are involved in sustainable agriculture, which restores the health of soil and develops a positive sense of belonging and community, as it also feeds people healthy food. In cities, walkable, bikeable, dense city centers, without cars, are allowing urban communities to develop in which people can have a deep sense of place and community. These local, networked healing practices can lead us all to find a sense of refuge in the places where we currently inhabit the world and can build community and sustainable relationships with the rest of the natural world.

What I like about Latour's analysis is its demand that we reorient politics to answer the question of how we all live well together in the matrix of crucial natural processes on which we depend. His view is that this alternative needs to be made as attractive as the fantasies of the two schools of thought that have been dominant until recently. On the right, it is a fantasy of going back to a sense of the local homogeneous community, buttressed by infinite exploitation of resources and border walls. Among globalizing liberals, it is the equally fantastical illusion of capitalist growth and progress that doesn't take nature, poverty, our need for a sense of place, or our interrelatedness seriously.

We need instead to imagine a sustainable world where the needs of all are taken into consideration, and where the natural limits of the biosphere are front and center in our planning. In that world, the Global North will take responsibility for the devastation of whole areas of the Global South and allow climate refugees to find new places to thrive and prosper.

We need to invest heavily in building communities that are socially and economically sustainable. The only ones who need to be worse off in that sustainable world are those who are profiting off our shared destruction. And if it is built through democratic public processes, this shows that we can work together to solve problems and work toward building that sustainable world and rebuild our common public.

Conclusion

In 2018 many people were jolted into action, or into higher level of action, by the IPCC report that the world must reduce emissions to net zero by 2050 and cut them in half by 2030. Some have argued that it is important to see the 2030 goal not as a cliff we are about to fall over but as a benchmark for the goal of total decarbonization by 2050. In other words, a lot of work has to happen right away. Much important work has been stimulated and given urgency by the twin 2030 and 2050 goals in that report.

Doing the work to achieve those goals will involve taking the political power away from the companies that have benefited, and continue to benefit from, the burning of fossil fuels and from consumerism. Building that world promises to lead to a better world for all of humanity and for the nonhuman species we share the planet with. Getting there as quickly as possible is the fight of our lifetime.

CHAPTER 2

Another World Is Possible

Imagine cities with public transportation so good we all preferred using it to driving.

Imagine quality health care that everyone receives as a right, so that people were not tied to bad jobs, or thrown into bankruptcy by medical bills.

Imagine affordable housing being built near shops and transportation, so people didn't need cars, didn't need to travel far to work, or to do the things needed to sustain their lives.

Imagine well-paying jobs for people from marginalized communities designing and building new green infrastructure to create a sustainable matrix for us all.

Imagine a country that supports sustainable farming, so rural communities would have lively farms in them, and people laboring on those farms wouldn't be poisoned by toxic pesticides.

Imagine an economy that was measured by how healthy it was for people and nature, using the Genuine Progress Indicator, rather than by how much was produced for sale, the GDP.

Imagine a society with enough equality and such cooperative systems for getting our needs met, that people only needed to work twenty hours per week to live well.

Imagine enough investment in education such that everyone could pursue the work of their dreams without going into debt, and people from all income groups could be prepared to work to build this better world.

Imagine living without a sense of guilt that every time we buy something, we are destroying the world, but not buying means someone loses their job.

At the present time there are over a million organizations working to make this transition happen, and their view of a sustainable world is remarkably consistent.[1] Transportation systems are being implemented all around the world which reduce emissions and traffic, while also making getting around less class divided. Renewable energy is now less expensive than fossil fuel energy. All around the world people are creating smart grids to distribute clean energy. Many countries have made tremendous progress in the problem of what to do with garbage. Money to finance new fossil fuel infrastructure is starting to dry up.

The things we need to do to make a just transition happen will almost all make for a better world, and in that sense, they come with little sacrifice. Sustainable farming practices mean that farmworkers and farm communities are not subjected to devastating pesticides, and there are more and better jobs in farming. Small-scale sustainable agriculture is much better for communities than industrial agriculture. People who eat diets lower in meat live longer than those who eat a lot of red meat.

All of these things are possible. And if we want to have a sustainable society, they are actually necessary. The technologies and know-how to achieve these things, within ten years, are already with us. With progressive taxation, the funding mechanisms are available. No one needs to suffer to make these things real. What is needed to get there is the political will, buttressed by the belief that it is possible. This chapter paints a picture of that better world and explores the importance of developing a vision of a just transition which is in contrast with a business-as-usual approach to getting off fossil fuels.

Believing in the Possibility of a Better World

When I first became interested in climate action it was hard for me to imagine how we could live sustainably. Would we need to give up on living in cities? Would we have to give up on pleasures like hot showers and cozy rooms? My inability to imagine the future made it hard for me to engage with climate work. What were we aiming at? What were the steps for getting there? Who or what were the targets of action for change? Was it politically possible to build the support to make the needed changes? I couldn't imagine a solution, and without that, it felt hard to know what the important steps were that I should take to deal with the problem. Without clarity on those questions, I felt stuck when I thought about taking action for the climate.

I was moved to action when I read George Monbiot's *Heat: How to Stop the Planet Burning.* That book, which was published in 2007, laid out some commonsense steps that would each lead to 90 percent emissions reductions in a variety of sectors of our lives. Written at a time when the cost of renewable energy was much higher than it is today, and many of the technological fixes which have become well understood were still in their nascent stages, Monbiot laid out a clear path to serious reductions with little social disruption.[2]

For me, what was so powerful was the thought that we could solve the climate crisis and still live lives of comfort and pleasure. That realization led to another: that if we could all live well in sustainable societies, then it might be possible to make the political case to get us there. Since 2007, as the crisis has gotten worse and emissions have risen, the solutions have only gotten better. And the clarity about what we are working toward has become more widespread.

Just Transition to a Sustainable Economy

Climate justice argues that climate crisis impacts different people differently, and increases existing inequities.

According to Maisa Rojas, scientific coordinator for the COP25 climate summit and director of Chile's Center for Climate and Resilience Research,

> We know the climate crisis acts as an amplifier of social inequality, disproportionately affecting the most vulnerable. Think for example about the heatwaves, of which we saw several this year. The impacts of such extreme events are experienced very differently depending on whether you have access to air conditioning or whether you have a park or greenspace close by. The ability to recover from the impact of a tropical storm also depends on your access to insurance or finance, and to natural resources such as water.[3]

The climate crisis is a threat multiplier. People of color, whose needs are not attended to by their government, will be worse off as the crisis develops. A government that doesn't care about the lives of low-income people and people of color will allow policies to develop that meet the needs of those who are closer to power. As sea levels rise, people with money will be able to relocate to higher ground and buy food as it becomes scarcer. Those living on the front lines of devastation will be those with less power: people of color and low-income people. The transition we are working for needs to be a just transition, which means one that takes everyone's needs into account and includes their active participation in crafting solutions.

Including the needs of people of color, workers, and people from low-income communities is not just an extra side issue thrown in to make us feel better about what we are doing. When people of color and low-income people are at the table when decision are made, it is much more likely that those decisions will help build a better world of all of us. People at the front lines of ecological and social devastation are the most likely to see the unintended, and sometimes intended, negative impacts of policies. A movement for a sustainable world needs to take power dynamics into consideration and make sure that policies aimed at addressing the climate crisis are developed with strong voices from the communities at the front lines of the devastation that is happening.

If labor unions are part of the solution, we are much more likely to get broad public support for those policies than if we stick with a probusiness transformation that removes fossil fuels but keeps exploiting labor. Unions are important because they allow working people a mechanism to fight for their fair share in a workplace. Where there aren't unions, corporations dominate, and when corporations have huge amounts of wealth and power, they can control political systems for their interests rather in the interests of the common good. Challenging economic and political inequality is a crucial part of the transition to a sustainable society.

A green economy can be developed in ways that offer well-paying jobs to people in communities that have been devastated by deindustrialization and to people from marginalized communities. If we win the fight for the right policies, those can be union jobs that pay decent wages and have good benefits. Creating broad support for such a massive social transformation requires that everyone—except the fossil fuel industry, the 1 percent, and the politicians who support them—is on board.

Business-as-Usual Environmentalism

There are many who argue that we can solve the climate crisis without disrupting the dominance of business. In *Climate of Hope*, authors Michael Bloomberg, billionaire and former mayor of New York, and Carl Pope of the Sierra Club argue that for many businesses it is actually a money saver to switch to green practices. They offer a vision of a win-win world, where business save money, as they innovate, and good public policy makes our cities more resilient and moves them away from dependence on fossil fuels.

Many of the things that Bloomberg and Pope call for are very positive, such as investing in energy saving in old buildings and planting mangroves in the Global South. There is much good that comes from people showing the ways that business can do well with sensible climate regulation. But what probusiness environmentalists rarely discuss is why it has been so hard to get people to implement those great money-saving innovations and the policies that nudge them into reality.

The work of a business-friendly transition can only go so far without a serious challenge to the power of businesses in several large-scale industries, such as fossil fuels, finance, and agribusiness, which are fighting to continue to run society in ways that gain them profits. That leaves us in a perpetual battle between government regulatory systems and the power of those with money. As long as those with a lot of money can buy politicians, can buy their way out of the negative impacts of the crisis, or have a vested interest in people buying disposable or unsustainable products, they will keep fighting to have the markets that create the conditions in which they make their profits, regulated to ensure profit, rather than run in ways that ensure the good of society as a whole.

Capitalism

An economy that puts the needs of capital first will always find it difficult to cut emissions. Capitalism, defined as an economy that allows private capital as much leeway as possible to pursue profit, allows social resources to be largely distributed on the basis of market mechanisms, with governments acting to provide the rules that determine how those markets work.

When housing is a private commodity and the development of housing is driven by the market, then it is very hard to get dense affordable housing near transit, which is the most sustainable form of housing. In the US, the average size of homes has doubled since the 1950s. This is not because we have bigger families. The opposite is true. Rather, it is because of people's desire for capitalist-driven status, and because most housing is built as a place for investors to put their money, and high-end housing is better at turning a profit than moderate housing. Only when public policy favors housing to meet human needs, rather than housing for the market, do we get housing that meets our social and environmental needs.

The climate justice framework asks us to always keep broader issues of power in our sight. It focuses our attention on the impacts of racism, inequality, and the operations of capitalism, on our understanding of what we are up against. One of the most prominent proponents of a climate justice framework is Canadian journalist Naomi Klein. In her book *This Changes Everything: Capitalism vs. the Climate*, she argues that dealing with the climate crisis is in alignment with a set of practices necessary for building a more socially just world. She argues that as she came to understand the things needed to solve the problem of climate change, she

> began to see all kinds of ways that climate change could become a catalyzing force for positive change—how it could be the best argument progressives have ever had to demand the rebuilding of local economies: to reclaim our democracies from corrosive corporate influence; to block harmful new free trade deals and rewrite old ones; to invest in starving public infrastructure like mass transit and affordable housing; to take back ownership of essential services like energy and water; to remake our sick agricultural system into something much healthier; to open borders to migrants whose displacement is linked to climate impacts; to finally respect indigenous land rights—all of which would help to end grotesque levels of inequality within our nations and between them.[4]

Capitalism began around five hundred years ago and was from the beginning linked with slavery and colonialism, which provided the investment capital that started the industrial revolution.[5] The power of capital, meaning money and assets used for the creation of more money, has spread. Along with that has spread ways of living that put profits before human and ecological needs. From its beginning, capitalism has been based on the exploitation of human and natural resources. Yet it is possible that this five-hundred-year trajectory is beginning to come to an end. It may be that the climate crisis will accelerate the push to build a better world. With the current urgent need to create economic and political systems which are resilient to climate disruptions, and which force us to make social decisions on some basis other than profit, there is an opening for more progressive alternatives to gain more traction than we have dreamed possible in recent decades.

One might think that if capitalism is the problem, then we are in even bigger trouble than we thought. We need to de-carbonize our economies in the next few years, and global anticapitalist revolution is likely to be farther off than that. Yet, as I argue in my book *Getting Past Capitalism: History, Vision, Hope*, the move away from capitalism can be a stepwise move, and the work we do to deal with the climate crisis can be important steps in that direction. There are many things we can do to help there be less capitalism. And maybe someday, after the urgent task for getting net emissions down to zero has been achieved, a more thorough routing of capitalism might be in order.[6]

Capitalism is not so much a system that needs to be overthrown as it is a set of highly destructive social practices, all of which can be challenged.[7] And it turns out that so many of the things we need to challenge to build a sustainable society are the same things needed to wean the world off capitalism.

One challenge to moving toward less capitalism is that once capitalist processes become a significant part of an economy and governments work in the interest of that aspect of the economy, people's well-being becomes increasingly dependent upon capitalists offering them jobs. And regulations to control the carbon emission of those companies come to be seen as risking "destroying the economy." Projects and policies that interfere with business as usual can result in companies closing, and jobs being lost, and that can lead to real suffering for the people who depend on those jobs.

As we work to build sustainable economies, we need to ensure that the policies we advocate for do not run afoul of the economic dependency trap of capitalism, where a society with a high level of capitalist practices causes people to become dependent upon the success of profit-driven business to survive, and so their support for it becomes a vicious circle. The section below explores some ways to break those dependencies. We need to increase ways people can meet their needs without relying on jobs in capitalist wage labor; we need to stop the forces that are driving the world to continue to invest in destructive practices such as creating new fossil fuel infrastructure; we need to use social resources to invest in the things we need for a just and sustainable society; and we need to change how we understand what is healthy for society and our own happiness.

Practices That Speed Our Momentum for a Just Transition

There are many things we can do to reduce the drag that capitalist practices and procapitalist ways of understanding the world have on our momentum toward a just transition. Some are huge and require social movements working over years, some are subtle and involve cultural change, and all are things already happening in the US and around the world. The following are some of the things which are important for building momentum toward a just transition. Some are obvious parts of climate action, but for others the connections are not as obvious.

1. Ditch the idea of growth and promote the use of alternative economic indicators

Economic growth, usually measured in gross domestic product (GDP) has come to be the most commonly used shorthand for describing the health of an economy. What it measures is how much is bought and sold in the capitalist market. The idea that an economy needs to grow for people to have good lives is one of the foundational myths of procapitalist ways of understanding the world. And it is devastating for the environment.

If I cook dinner for friends, that does not count as economic activity in the GDP. If I buy dinner at a restaurant, it counts. Measuring an economy based on GDP encourages us to think that more buying and selling is in itself a good thing. It discourages us from asking a deeper question: what we are doing that helps us meet real human needs and develop environmentally sustainable practices.

For human beings to survive, we do need to produce food and other material goods. We also need to engage in domestic work such as taking care of children and preparing meals. Feminist economists have argued that much of what we do to meet our needs through household labor should be considered productive activity, and should be considered economic activity. Yet that activity is invisible to standard economic measures such as GDP.[8] A world where more of that work was done outside the world of buying and selling is likely to be more satisfying than one where all labor becomes capitalist wage labor. The meaningful labor we do to meet our needs is qualitatively different from the, often meaningless, labor, which Marxists call "alienated labor" associated with jobs in capitalist production.[9]

If our economies are measured only by GDP, we will have the false belief that we need more capitalist production to have good lives. If people are told every day in the news that the cause of their poverty is a lack of economic growth, they will support progrowth policies. We need to help people distinguish what promotes social well-being from what promotes capitalist growth. Sometimes growth leads to more jobs and sometimes it doesn't. But increases in well-being are always good.

One important way to wean people away from a belief that economic growth is always better is to promote the use of better economic indicators. If we measure an economy based on things directly associated with well-being—like literacy, longevity, and reported happiness—then we will be able to figure out which policies are good for "the economy," and distinguish them from policies that increase capitalist economic activity, but may not improve well-being. There are many alternative economic indicators that economists working for a just transition have developed, such as the Genuine Progress Indicator (GPI) and the Human Development Index (HDI). Governments all around the world are beginning to use them.

2. Reduce work time

Measuring an economy's health based on GDP can lead us to believe that more work is a good thing, even as most people don't like their jobs. Yet most people don't like going to work, and many jobs fuel the destruction of the planet, and help corporations rake in the profits that they use to manipulate our political systems.

The oddity of this is apparent in the following two hypothetical scenarios: First, imagine a society in which the average workweek is eighty hours. Workers in this society would pay others to care for their children and elders, buy premade food and consume commercial culture for entertainment. They would experience high levels of stress, and have high carbon footprints.

Next, imagine an alternative scenario in which everyone works twenty hours a week. In this scenario, people have time for friendship and time to care for children and elders. They have time to enjoy making things and cooking for themselves and for each other. Most likely they are happier and have lower carbon footprints, as they don't need to commute as much and they buy fewer disposable products. By spending more time at home, they would live in communities with stronger social fabrics, and would be able to develop forms of pleasure that are not related to commercial consumption. The first scenario is one that would have a GDP much higher than the second.

Struggles for shorter workdays can have huge social and environmental benefits. The move in the industrialized countries to the eight-hour day led to major improvements in well-being. France's move in 2000 to a thirty-five-hour work week, without a reduction in pay, was a major advance. Shorter work hours have a variety of benefits. The individual spends less time in alienated labor and so has a better life. There is more possibility for gender equality, as there is less time stress in families around who can take care of children and elders. It also leads to lower unemployment, as employers are nudged to hire someone else for the other hours.

A big part of a just transition is minimizing people's dependence on waged labor and bringing pleasure and meaning back to other forms of work that take place outside of wage labor. This can be done by creating meaningful jobs that people enjoy and by promoting policies that allow people to perform waged labor for fewer hours each day. It can also be promoted through the creation of worker-owner coopera-tives, which enable laborers to have power over their work processes and to control the profits of their labor.

3. Strengthen social safety nets

As we work toward a just transition, many jobs in the current economy will no longer exist. The more we have social systems in place to take care of everyone's needs, the less trauma people will experience in that transition. For many people, a big part of what is devastating about losing a job is that the job is what gives them access to health care for themselves and their families. If we had a strong national health care system, people wouldn't need to work at a job to have access to health care. Similarly, when we don't have good retirement systems, people need to keep working until they are very old, and aren't able to retire with enough money to be economically secure.

There are many countries in the world where the safety nets are strong enough that disruptions caused by losing a job are not devastating. This is true in wealthy social democratic countries, such as Sweden, where the wealthy and corporations are taxed highly enough to pay for a strong social safety net. It is true in socialist Cuba, where the government pays for health, education, and retirement. It is also true in very low-income countries such as Bhutan, which have strong social bonds and low levels of economic activity but high levels of safety and satisfaction with life.

4. Develop community capital

One reason that capitalist economies require growth is that investors will only invest if they believe they will get more than they put into a given project after some period of time. This aspect of a capitalist economy is like a Ponzi scheme. Many players can do well if the pie keeps growing. When an economy contracts, investors at the bottom of the pyramid can lose everything. As long as the economy grows, however, most investors improve their situations.

One of the most crucial aspect of a just transition is stopping the money that is financing the fossil fuel industry. As of this writing, banks, pension funds, and other large investors are still putting money into developing new fossil fuel infrastructure. A new fossil fuel pipeline is intended to last around fifty years. As a result, those investing in that infrastructure then become committed to keeping the extraction and burning of fossil fuels going. A crucial aspect of the movement for a just transition is to stop the flow of investment capital to those projects.

As finance capital has come to dominate over other forms of capital all around the world, in recent years, we have seen increasingly dysfunctional demands for ever-quicker returns, as productive companies are devastated when they are bought by investors looking to get quick returns. A community that is under-resourced will go to great lengths to entice companies to locate there, even if the companies offer very little to the community. The mere hope of jobs and a small amount of tax revenue is often enough.

This doesn't mean that a society can live without investments or that investment is necessarily a bad thing. Investment pays for things that we won't see the rewards from for a while. Farmers need investment capital to buy inputs for a crop that they won't be able to sell for a while. Producers of shoes need to invest in equipment and materials. One challenge for an economy not based on growth is figuring out how to obtain the investments that are needed for economic activity. We can do without much of what happens in a capitalist economy, such as advertising, the production of disposable or superfluous goods, and the sales and fees associated with for-profit health insurance. Yet many forms of economic activity are crucial for good lives. In any economic system, investment is necessary.

But what if governments or local communities had their own capital to invest in socially useful projects? In contrast to finance capital, which is always looking for a short-term return, or venture capital, which is looking for the next big thing to generate profits, community-based "patient capital" can be invested in building a sustainable economy. The movement for public banks is an important part of this, as governments can mobilize huge amounts of capital that doesn't need a quick return.

Investment in a healthy economy requires "patient capital," which asks for a return, but a reasonable one, and over a fairly longtime horizon. Governments at any level can leverage the money they have on hand, in things like retirement systems, to create public banks that are tasked with investing for the public good. Nonprofit institutions can be formed to share the capital created by worker-owned cooperatives to invest when one cooperative has more than it needs at the moment. To the extent that an economy has a noncapitalist sector, there are many ways to get the resources needed to invest in public goods that don't rely on growth as an incentive.

5. Wean ourselves from consumer culture

Given the billions of dollars spent on advertising and in commercial media to promote a high-consumption lifestyle, it isn't surprising that many people think that the path to happiness is in ever-higher levels of consumption. But if only some people are able to experience those high levels of consumption, the total level of happiness in the society goes down. Empirical studies of happiness find that inequality breeds unhappiness, as people experience status anxiety and as the social fabric becomes torn apart.[10]

If people feel that their sense of themselves as a successful person is dependent upon the forms of status that the advertisers have convinced us are crucial, then we will think that huge houses are better than modest ones. We will be attracted to flying all over the world for vacations we can share on Instagram, rather than spending time exploring our local surroundings. And we will need to buy throwaway fashions to keep up with the latest trends.

High levels of social solidarity and connection are far more important for happiness than high levels of consumption. In order to be happy, people need a sense of security, access to food and other basic needs, and a sense of community. Working to challenge the cultural forces that lead people to believe consumption is the path to happiness is an important part of a move to a sustainable economy. This is not an easy task, but our culture is shifting around an understanding the unsatisfying nature of conspicuous consumption.

6. Promote a sense of sufficiency

One thing that keeps us on the hedonic treadmill of thinking we need to produce and consume more is the belief that we live in a world of scarcity. This view is promoted as a basic fact in most economics classes. We are told that Economics is the study of scarcity and that there are hard questions in society about how we allocate scarce resources. The truth is that there is enough in the world to satisfy everyone's needs. The world produces enough food to feed everyone well, the problem is in how it is distributed. There are enough homes in the US for everyone to live in, the problem is how they are distributed. There is enough clothing produced in the world for everyone to have good clothing, the problem is that we are convinced that we need more, and newer. Poverty, and a lack of sufficient food, clothing, and housing and are not caused by there not being enough stuff. It is caused by the stuff not being allocated fairly.

In the US we have incredibly unequal distribution of wealth. If we were to have a progressive taxation system, which used to exist in the US, and still exists in many countries, we could use that money to fund social goods. Less severe than wealth inequality, but still very severe, is income inequality. To show that there is enough income in the country for everyone to live well, Gar Alperovitz engaged in a simple mathematical demonstration. Look up the number of people in the US. Look up the total national income. Divide it and clump it into families of four. If we had completely equal levels of income in the United States today, every family of four would make $200,000 per year.[11] If the two adults in that hypothetical family were to work half-time, the family could have $100,000 per year. There is enough income in this country for everyone to live well. The problem is inequality, not a lack of sufficient resources.

In the middle of the twentieth century, when the working class was strong in the US, unions fought for a larger share of what a company gained to be paid out in higher wages. Since the 1980s, the US government has been hostile to unions and has made it harder for them to organize. As a result, the working class has been less powerful, and wages have stagnated. Yet in that same period, the value of what companies have produced has continued to grow. If, instead of going into profits, the benefits of increased productivity were to go into work-time reduction, then every twenty-four years we could cut our work time in half, and still have the same incomes we did at the start. If we had started that in the 1990s, then full-time workers could all be making their present salary while working twenty-hour weeks.

There is plenty in the world. And there is plenty of money to invest in a just transition. What is keeping people in poverty and keeping money flowing to invest in the fossil fuel industry, are the results of the ways that procapitalist thinking encourages us to allow people with an interest in profit to make decisions about how our resources are invested and distributed. We can address these problems by promoting progressive taxation and minimum-wage increases. More effective labor struggles, too, result in a larger percentage of income going to workers.

7. Embed markets in intentional social processes

A society with markets is one where people can buy and sell, and how much they want a product is part to what sets the price for that product. There is nothing inherently wrong with markets, but under capitalism we are told that markets need to be "free" meaning that we should not subject them to social decisions. Yet there is no such thing a as free market. Markets are always created by social decisions.[12] The idea of "free markets" is a place where much manipulation in the interests of the wealthy takes place.

That basic procapitalist idea is that when there is demand, some businessperson will be incentivized to supply that demand. The problem is that people without money do not have demand that matters to a businessperson. If I am hungry and have no money, no business is incentivized to meet my demand for food. "Effective demand" is demand backed by money. That is the kind of demand that stimulates business.

Procapitalist thinkers argue that things like the minimum wage "distort' the natural market in labor. Yet, because owners of the means of production have more power than workers, without a minimum wage law, wages can go so low that people will not earn enough to survive on. Over the past hundred years, all around the world people have fought for a minimum legal wage, so that those who work can survive, and maybe even thrive.

One of the problems we face in the current period is that many people believe that the forms of regulation we need to make a just transit to a sustainable society are "bad for the economy" or will "distort markets." Yet the move to renewable energy happened because governments invested in the research needed to develop them, and they subsidized their early development before they were economically competitive in a capitalist market.

Markets are fine ways to distribute goods and get a sense of what people want. But they only work when they are a limited part of economic decision making. They can't be the main force in deciding where investments go and what the rules are under which things are produced and distributed.[13]

8. Challenge ruling-class actions

In the present period there are major transnational structures, such as the International Monetary Fund, and the World Bank, as well as transnational trade deals, which function in ways that favor the interests of large-scale capital. The people who meet in Davos, Switzerland, every year for the World Economic Forum to discuss the state of the world, have been generally committed to keeping the global economy functioning for economic growth and the generation of profits. They have been less concerned with environmental sustainability and even less concerned with equity.

A significant part of the work of a just transition needs to involve struggles over those trade deals and the rules that are then embedded in markets and laws that favor unsustainable and exploitative practices.

Conclusion

As long as growth and employment rates are the measures of a healthy economy, our environmental interests will be at war with our economic interests, and what is good for capital will be seen as what is good for everyone. We need to wean ourselves off the belief that increased GDP and more work are crucial to improving people's well-being. But we must also transform society such that building sustainability projects can be a real way for people to meet their needs. For that to happen, sustainability projects need to be part of a broader movement to lessen the economic dependency trap of capitalism.

The work we need to do to deal with the climate crisis requires an end to neoliberal trade deals, and politics driven in the interest of fossil fuel companies. It also requires that we build sustainable cities and that investments in things like housing, transportation, and health be driven by the imperative to satisfy human needs, rather than by the imperatives of capital.

In a society that says we need to have more and bigger to be social successes, it is hard to see a path to happiness that is sustainable. One of the things needed in transforming to a sustainable society are ways to connect with each other and our communities that give us a sense of meaning and purpose and social success that is delinked from consumerism. And it requires that we learn to develop forms of satisfaction that are grounded in the development of resilient communities, rather than through the chase after consumer pleasures.

Some of the most far-reaching proposals for dealing with the climate crisis include universal health care, and end to student debt, support for indigenous communities, reforms to farm labor laws, fights against inequality, support for organized labor, and moved toward racial equity in job opportunities. Those "other" issues are included because of the ways that these problems are interrelated. People who are insecure in their jobs will continue to support carbon intensive job growth. People without health care, or a reliable retirement system, will remain stuck in jobs, even if those jobs are unsustainable.

If farming in unsustainable ways is more profitable than sustainable farming, it will likely remain unsustainable, farm work will continue to be underpaid, and farmworkers will continue to suffer from pesticide exposure. If you change the laws around farming to prohibit the use of chemical inputs that are harmful to workers, then the farm economy becomes a healthier economy. Indebted students are much more likely to need to work in an unsustainable industry than young people who can follow their dreams and work for lower wages doing meaningful work.

The way to get people on board with the transition to a sustainable society is to make sure that their needs are front and center in the transition. We can wean ourselves from fossil fuels, and achieve zero greenhouse gas emissions, while simultaneously strengthening our communities, eliminating poverty, and having more joyous lives. The work we do to build a sustainable society can also be the work needed to reweave our social fabric in ways that create the homes, sense of community, and sense of purpose and success in life that lead to happiness.

CHAPTER 3

Messaging That Encourages Action

" If people don't wake up and stop doing all of the stupid things they are doing to destroy the planet, like buying plastic water bottles and eating too much meat, we are all doomed. People need to go green or the world is going to end. For all of us who love the outdoors, we need to think about the nature we love and how we are killing it. I can't believe that no one is doing anything. Well, at least I can do the things that need to be done in my own life.[1]

" Big changes need to happen, like retrofitting old buildings, developing public transportation, and switching to renewable energy. The good news is that those things are happening. We need to educate people about how important they are.

" I hate to break it to you, but the time for action to reduce emissions is over. We've already passed the point of no return. At this point we need to focus on developing resiliency in our cities and agriculture to survive the coming apocalypse.

" The earth, our home, is the only place we have to live. It's the place where we live in interdependence on bugs and birds and trees and mosses, where we live in deep interdependence with all other human beings. And right now it needs our help. Each of us can work to stop the forward motion of that machine that puts profits over the needs of the web of life that we are all a part of. Restoring the health and resilience of that web is the most urgent task before all of us, and each of us can work to repair it.

> Burning fossil fuels and destroying rainforests needs to be stopped as quickly as possible. We are already experiencing devastating wildfires, hurricanes with increased intensity and the arctic sea ice is melting at an astonishing rate. If we keep allowing those things to happen, our planet will become uninhabitable for our species, and for many others, as our atmosphere is destroyed. People of color, low-income people, and anyone without political clout are going to be hurt more than those with political power. Burning fossil fuels causes asthma and heart disease, and those impacts are much worse in low-income communities and communities of color. There are about twenty companies responsible for most of the damage, and those companies have tremendous political power. We need to do all we can to take away the power of those companies to destroy our shared home. As we get off fossil fuels, and take control over our economies away from those who put profit over our well-being, we can build a better world for everyone. Millions of people around the world are involved in that transition, and being a part of that movement can be inspiring and reduce your anxiety about the climate as you build that better world.

Narrating Change

As we speak with others about the climate crisis, we need to be strategic in our messaging. We need to talk about it in ways that resonate across all racial demographics, and that show the problem to be happening now, impacting people and their loved ones, and solvable by moving to a better world. And we need to be clear about who and what is blocking progress toward to a sustainable society.

The book *Re:Imagining Change: How to Use Story-Based Strategy to Win Campaigns, Build Movements, and Change the World* argues that the stories we tell have a huge impact on our ability to make a difference and that stories take place within a narrative frame, which sets the rules and expectations for that story.[2] One of the hard things about working on climate change for many years was that that most people saw it as something that was both everywhere and nowhere. In that narrative frame, almost everything we do was seen as causing the problem. All of us were seen as equally responsible for causing the problem and for solving it. The solutions were the small choices we could make as individuals.

While it might seem like it is "empowering" to feel like we can take the problem into our own hands as individuals and act differently to solve a really serious problem, that way of framing the problem has had terrible consequences. It structures the story of the climate crisis within the narrative frame of individualism, and the good people who take action versus the bad people who don't change their personal habits. Because that storyline implies that we are all equally the problem, they we all feel a little bit guilty.

One person making a different individual choice only really matters if others are inspired to follow suit. It is hard with a guilt-based message to get many others to follow suit. And, on the other hand, when we analyze what we are up against, the things that need to happen, at the speed they need to happen, we see that those changes are much larger than the aggregation of small individual choices. That standard story of collective responsibility erases the lines of responsibility that need to be focused on to understand where to put pressure to force rapid change in how our societies function.

The reality is that change is being slowed by a small number of large companies who have a vested interest in business as usual. It is important that we tell that story and focus people's attention on the kinds of things that need to be done to stop those companies from being able to slow the transition. Focusing on those companies and the politicians who support them puts the attention on the places where the biggest roadblocks to rapid action are right now. Climate activists have more recently begun to focus their messaging much more on the small group of large companies that are most to blame for causing the problem, on the banks that fund their development of infrastructure, and on the politicians who enable them.

It is generally a good idea to avoid polarizing in ways that create a small group of virtuous climate activists and the bad people who aren't acting. And it is generally a good idea to not dehumanize other people in our work for a better world. But it is important to be clear about who and what is blocking progress. People who focus on story-based strategy argue that a good villain is a crucial part of a compelling narrative. The fossil fuel companies, large-scale agribusiness, the banks who fund them, and the politicians who do their bidding are good villains in our climate justice narrative. That is not to say that the people who do that work are inherently evil, or that their humanity is irredeemable. But it does say that the choices they make, and the corporate charters they operate under are in fact the main things keeping us on a path of destruction. And that makes them the villains of this story.

Race and Climate Messaging

Climate messaging, in the US at least, needs to take racial differences seriously. Surveys have shown that there is more support for strong climate action among people of color in the US than among whites.[3] At a global level, the movement is in many ways led by people of color from the Global South. And there are stereotypes around climate action that it is a white enterprise. In reality, the movement is deeply multiracial, and those stereotypes keep it from becoming more so.

The mainstream media and some organizations persist in putting out an image of a mostly white movement. Vanessa Nakate is a Ugandan youth climate activist who founded the Youth for Future Africa and the Rise Up Movement. She gained worldwide attention when a widely circulated photo of youth climate activists showed four white girls she was posing with, and cropped her out.

While the media attention given to Greta Thunberg's strike in front of the Swedish parliament was an important catalyzing moment, the youth in the movement, while embracing Greta, have also taken every opportunity to stress the roles of other youth, particularly youth of color, many of whom were doing that work before her. Xiuhtezcatl Martinez, founder of Earth Guardians, is a hip-hop artist who gave his first speech at a transnational gathering in 2006, at the age of six.

For white people in the movement it is important to be aware of those dynamics that tend to give louder voice and attention to us, while also using our privilege to shift the conversation toward the voices and perspectives of people of color. It is crucial that white activists do not shy away from doing important climate justice work because they are afraid of taking up too much space in the movement. The movement needs all of us to do as much as we can. And white allies who understand the nature of power and privilege can play important roles disrupting the racism inherent in all aspects of our societies.

For the whiter parts of the movement in the US to be welcoming to people of color, it is important that the framing around calls to action on the climate crisis take seriously the pitfalls of unconsciously racialized narrative frames. Focus on saving nature as something separate from our selves is not generally a racially inclusive narrative, since it ignores the ways that we are all dependent upon natural process in our everyday lives, and frames nature as something to be visited and enjoyed. Messages that focus on local and health-related impacts of climate disruption and bad air are much more amenable to racially inclusive framing.

Why the Human Brain Is Not Well Adapted to Dealing with This Crisis

In his book *What We Think about When We Try Not to Think about Global Warming*, Norwegian psychologist and economist, Per Espen Stoknes argues that climate change activists need to take seriously the ways the human brain processes information in order to get people on board to make the kind of changes urgently needed to address the climate crisis.

He shares the results of psychological studies on how the human brain tends to work, so that we can build our actions in ways most likely to succeed in motivating people as they are, and not as we wish them to be. The book develops powerful messaging strategies, also based on research-tested approaches to messaging and action.

Climate communicators need to help people to see that the climate crisis will impact them and the people they care about. They need to find ways to shape the culture such that doing the right things for the climate give social status. They need to give a sense that many people are concerned and acting. They need to highlight short-term negative impacts such as fires and floods. And they need to help people to see that the risks are severe, but with strong action, can be mitigated.

Stoknes draws on insights from cognitive psychology to argue that we need to stop looking at the climate crisis through the frame of loss and sacrifice. Otherwise, it will be hard to gain traction. People hate losses more than they love gains. Helping people to see how much money they will save with better insulation and solar energy is a more motivating frame than telling them that the path to success is to give up on the pleasures of a consumer lifestyle.

The other important concept he draws from cognitive psychology is cognitive dissonance. Many people hear the bleak messages we put out as accusations. They want to think that they are good people, and if we say that good people don't live the way the rest of society tells them they should live, then they will resolve the dissonance by distancing themselves from the truth about the climate crisis messages they hear. The power of denial is so strong that people generally resolve cognitive dissonance by changing what they believe, rather than changing how they act.

How we think about a claim is deeply impacted by the social circles we run in and how our associates see the world. And that gets us to Stoknes's insights from the psychology of identity. Here is the most stunning finding from the research he reports on: "The better science literacy someone holding a conservative ideology has, the more wrong, she will be on the climate science."[4] Educated conservatives have the facts, but they live in social groups whose identities are invested in not believing in them. The more you throw facts at them, the more activated their systems of denial become.

Stoknes tells the story of a white South African woman who changed her view of apartheid when someone working in her home was almost killed. The change started with the emotional, and then worked toward the cognitive.

> Such a break always involves emotions that bring about a shift in the organization of personality, and thus a subtle shifting of identity. . . . What is needed is the work of a cultural movement similar to the ones that dismantled apartheid, abolished slavery, or took on nuclear arms. What remains unpredictable is when this steady, seemingly ineffectual, exhausting work will result—or coevolve—into a seismic shift in cultural denial. But one day after the swerve, what used to be resisted and denied becomes a new, shared reality.[5]

No matter what we do, there will be a sizable proportion of the population who will never be with us. The good news is that a just transition to a sustainable society does not require that everyone be on board. We just need enough support from people doing the right things to change social policies.

Stoknes summarizes his discussion of denial with the five walls that make it hard for climate messages to hit their target:

- Distance, the issue seems remote;
- Doom, if it is seen as a disaster that requires sacrifice, it creates a wish to avoid;
- Dissonance, if what we know conflicts with what we do, then we will try to not know it;
- Denial, if we can deny that it is happening, we can continue to feel good about ourselves;
- iDentity, "cultural identity overrides the facts."[6]

He then argues that we need to acknowledge the realities of hard psychology we are up against. The fossil fuel industry has done a great job using all of these aspects of human psychology to lull people into inaction. We need to "move more with the flow of the human psyche. . . . People want to live in a climate-friendly society because they see it as better, not because they get scared or instructed into it."[7]

He proposes the following to deal with these limitations: Make the issue feel near, human, personal, and urgent. Use supportive framings that do not backfire by creating negative feelings. Reduce dissonance by providing opportunities for consistent and visible action. Avoid triggering the emotional need for denial through fear, guilt, self-protection. Reduce unnecessary cultural and political polarization.

Strategies for Effective Climate Communication

Given those realities of the human brain, and the complexities of what we are up against, Stoknes offers five major strategies for effective climate communication:

1. Social. "Since the imitation effect is so strong, asking consumers to go green will fail unless they are convinced that many others will do the same." Rather than telling people that it is terrible that no one is doing anything, remind them that there are over a million organizations in the world right now working to make the transition happen. They can be part of that big team.

2. Supportive. Employ frames that support the message with positive emotions. While the news media survives on sensationalism, a sense of despair and fear rarely motivates action. Instead we need to talk about climate in terms of "insurance, health, security, preparedness, and most of all, opportunity." Climate action is motivated by a sense that one is doing something positive.

3. Simple. Make climate friendly behaviors easy and convenient. We need to develop policies that make green behavior not too hard and then nudge people toward those better behaviors. Most of our efforts should be aimed at policy change, and very little at asking people to live differently than the people around them. Universities are now eliminating food trays so that students don't take too much food and waste so much. These changes require hard work at policy change, but do not ask consumers to make heroic leaps in their everyday choices.

4. Story-based. Use the power of stories to create meaning and community. One important force for generating climate action is the narrative that we're building a beautiful, sustainable world, in which people will live better than we do now. Sustainable walkable cities are more fun to be in. Eliminating the burning of fossil fuels makes the air better to breathe and healthier. A more plant-based diet can be more delicious and help us to live longer and feel better. Places with more trees are nicer to be in. A world with less consumerism is a world with more community, sense of purpose, and happiness.

5. Signals. Use indicators for feedback on societal response. In order to maintain engagement in societal transformation, there has to be a way to speak of and get feedback on any actions that turn us in the right direction. Without such feedback, there is little learning and less motivation—only increasing confusion and helplessness. Thus, we should push governments to switch to using alternative economic indicators, such as the Genuine Progress Indicator (GPI) to replace GDP, by which one measures how much stuff is produced and sold but not what the social impacts are of that economic activity.

Children, Youth, and Climate Messaging

For many years, people doing work on climate change have focused on the idea that we need to save the world for our children. By focusing on one's own children, one thinks of the world in terms of protecting my own, which for people with more privilege can mean promoting policies that are good for my family but not necessarily the families of others. It is also a message that rankles people who don't have children, and can center the heteronormative family. Rather that pushing our consciousness to focus on the broad social fabric of our interconnections, it nudges us toward private and individualistic solutions.

As the climate youth movement has become more powerful and widely recognized, it has been interesting to see how the young themselves have framed their messages. They don't ask their parents to protect them. Rather, they make their claims as full participants in society who don't have a lot institutional power, demanding that those with that power use it to achieve the rapid social transformations that are needed. They tend to focus on the need for adults to take responsibility to challenge power and shift systems to meet all of our needs for a livable future.

Conclusion

In order to make our messaging as impactful as possible, we need to stop being that person at a fire who stands around screaming for people to do something because the situation is urgent and no one is doing anything. Instead, we need to paint a picture of an inclusive sustainable world we are trying to build; remind people that the work is already underway and that millions of people are doing important work. We need to paint a picture of a movement that takes a wide range of people's concerns seriously, promises solutions to those concerns, and, most importantly, gives people pathways to significant effective action. And it wouldn't hurt if getting involved was shown to be fun and able to bring people into community with other people who are caring and inspiring and who find their climate activism to be an enriching part of their lives.

CHAPTER 4

The Large-Scale Solutions

We are all increasingly hearing about the terrible droughts, wildfires, hurricanes, and other disasters caused by the climate crisis. Activists have rightly focused people's attention on the urgency and scale of social change that needs to happen to meet the 2030 and 2050 goals set out by the IPCC. When I talk to people who are not involved in climate crisis work, so many of them ask "when will people start to do something?" People outside the climate movement hear very little in mainstream news, or even in alternative news, about the things that are being done to build a sustainable world.

This is a problem, because, as we saw in chapter 3, people tend to take their signals on how to act from those around them. If no one else is doing anything, then I may grieve on my own but may not be pulled to do anything either. But it turns out that huge things are being done, all over the world, and much of it at the massive scale needed to make the transition we are working on. And there is work every one of us can do to accelerate the pace of adoption of those sustainable practices. Knowing those examples and sharing them is an important part of climate activism. It gives realistic hope and keeps us focused on our goals.

One of the noteworthy things about the time were living in, which is both terrifying and encouraging, is that so much of what has been stable in human societies is coming unglued, and societies are transforming at a remarkable pace. With the internet and other forms of sharing, it is increasingly the case that when a good practice is developed in one place it can move rapidly and be adopted all around the world in increasingly short time frames. What is keeping the change from happening at the pace that is necessary to keep us from burning down the home we call earth, is the power of entrenched economic interests and their control over political systems.

As we do our work to shift the political calculus of the people who make major decisions which govern how our societies operate, amazing transformations can happen fairly quickly. This chapter describes some of the major things that need to happen to get us to net zero emissions as soon as possible. Some of the more exciting and far-reaching transformations are well on their way. Keeping clear about what needs to happen and the ways that much of it is already happening, and holding those realities in our minds and our hearts can give us the courage to keep fighting. And sharing these ideas can help others step out of their climate despair and join the movement.

This chapter focuses on big things, but one important thing to remember is that big things often start out with small pilots. In 2019 the City of Berkeley, California, outlawed gas hookups in new buildings, since burning natural gas in our homes is a significant source of greenhouse gas emissions. It turns out that for a new building it is cheaper to start without gas. People working on that law needed to figure out all sorts of picky technical issues. There is not a lot of new building happening in Berkeley, so this might seem like a lot of work for a small gain. But now that all of that technical and political work has been done in Berkeley to make that proposal viable, it is very easy for other small towns, big cities, states, and even countries to adopt a similar measure. And that spread is already happening.

Sometimes when I sit in meetings, working on something small like that, I feel overwhelmed by the smallness of what I am doing. I hope this book encourages you to think big and audaciously. But it is also important to not underestimate the power of the small things we do to create models, and to do your part for those small things that might add up to the large-scale rapid transformations we need.

We should always ask how our small work might add momentum to larger changes, and put our energies where they are likely to have the most impact, knowing full well that we can never have a really clear answer to that question. The large-scale changes that are needed for a just transition to a sustainable society are underway, are exciting, and many of them are being adopted widely. It is our job to increase the speed of their adoption. For so many of these large-scale systemic transformations, the engineering is already there, and almost all of the innovations being adopted are cheaper and more efficient and lead to better social outcomes, than the current destructive systems in place now. What is required to make them happen is political will. And increasing political will is the work of climate justice activism.

Many of the examples on this chapter come from the book *Drawdown: The Most Comprehensive Plan Ever Proposed to Reverse Global Warming*. Edited by Paul Hawken, it is the result of Project Drawdown, a team of analysts from all around the world who came together in 2013 to study and rank the one hundred most significant things needed to address the climate crisis. What I like about the book is its scale and how specific it is about the potential impacts of each of the solutions it analyzes. A lot of very serious work by a global team of scholars went into analyzing where each solution is and its potential for even greater significance. One shortcoming of the book is that it doesn't call out consumerism as a separate category. Many analysts believe that we need to get past a society based on consumption as a pastime and path to status, and goods that are made to be thrown away and replaced quickly. That issue will be included in the discussion below.[1]

The Exponential Roadmap is a similar project, written by a team of experts, that gives a comprehensive overview of what needs to be done by 2030 and 2050 to achieve a transition to a sustainable society. That analysis focuses on many positive changes which are happening at exponential rates, and it makes the case for what needs to be done to achieve those goals.[2] This chapter relies on that analysis for details on some of the hopeful large-scale transformations that are underway.

The chapter pulls out a few of the major categories that this work falls into to give a quick story that is relatively easy to tell about what has to happen, and what is happening, in each sector. It focuses on systems of reducing carbon emissions through transforming energy, transportation, buildings, consumer products, and agriculture. It then looks at how greenhouse gas levels can be brought down by planting massive numbers of trees and sequestering carbon in the soil. It looks at how ending global poverty is an important part of a just transition to a sustainable society. Finally, it considers best practices in adaptation or ways to make our world more resilient to sea level rise, storms, fire, and drought.

Energy Sector

The key to averting the worst of the climate crisis, is to stop the use of fossil fuels as rapidly as possible. Energy is one of the most crucial, as well as one of the most rapidly changing aspects of the global economy. If everything switches to electricity and electricity is sustainable and affordable, that goes a long way toward making a just transition away from fossil fuels. Already wind and solar are generally less expensive than fossil fuels.

In 2015, the US government spent $649 billion to subsidize fossil fuels. We are still digging the pit we need to get ourselves out of. Making the energy transition involves ending the subsidies to fossil fuels, investing massively in developing more renewable energy, and changing our energy grids to be more efficient and resilient. Working to stop the fossil fuels subsidies is critical. As long as fossil fuel companies have a stranglehold on politics, they will keep working to slow down the transition. And fighting against subsides is an important way to expose the depth of that stranglehold.[3]

Many organizations are working to stop the development of new fossil fuel infrastructure, such as pipelines. That is especially important because once the money has been spent for something like a pipeline, it is then cheaper to keep it running than to lose all the money invested and stop using it. Fights against new fossil fuel infrastructure and against subsidies to that industry are difficult political fights. All around the world indigenous people have been at the front lines of those struggles.

Less of a fight but also important, is the development of renewable energy. Wind energy, solar farms, and rooftop solar together are huge aspects of the transition we are experiencing. Because they are now economically competitive with fossil fuels, it is no longer such a big political fight to get them developed. The financing mechanisms for rooftop solar still need to be pushed through slow political systems. There are debates about the best ways to develop solar energy, with some arguing that it is more efficient to develop large-scale solar farms, and others arguing that there are more benefits and fewer negative environmental impacts from rooftop solar. Similarly, there is work still being done to figure out how to mitigate the impacts of wind power on migratory birds.

Once electricity is generated, it needs to be bought by a power company and then either stored and distributed or just directly distributed. I live in Northern California, where our electricity and gas have been supplied, since 1905, by the private for-profit company Pacific Gas and Electric (PG&E). PG&E, like many energy companies, is the worst of capitalism and state socialism. It is a monopoly, so it isn't pushed to compete and improve, which is one of the only virtues of capitalism. But as a for-profit company it must make decisions on how it operates based on what will turn a profit for shareholders, so it has mostly not invested in new technology or in maintenance. This is one of the reasons that, as California has become drier as a result of climate change, it then experienced devastating wildfires, such as the ones that destroyed a lot of the town of Santa Rosa in 2017 and most of the town of Paradise in 2018. PG&E was more interested in profits than in investing in safety. So, when the state experienced hot dry winds, electricity infrastructure sparked wildfires.

In the US, electricity infrastructure was developed at the beginning of the twentieth century with a lot of support from public entities, doing research, setting standards, and investing in the first electric grids, the large-scale system through which energy is stored and distributed. After that original investment of public resources, the grids ended up being mostly held in private for-profit hands, but in many cases as protected monopolies. At this point in the US, grids are so poorly maintained and designed that they waste 65 percent of the energy put into them.[4]

A crucial part of the clean energy transition is to develop grids that are safer and more efficient. One way to do that is to take back the grids to public control. If a grid is run by a well-managed public entity, it is more likely to serve the public interest that if it is run by a for profit corporation.

In much of the world, investments need to be made in more modern efficient and safe ways to transmit electricity. Microgrids draw energy produced in a small locality such as a small town and distribute what is produced locally back out to the local community. They are especially effective in the large parts of world that have no electricity infrastructure at this point. A town can have its own rooftop solar or wind turbines. It can link those with battery infrastructure and local transmission lines to serve a small area. These microgrids then can be connected to larger grids, in case there are local short-term shortages, or there is extra capacity that can be sold to the larger system.

In addition to microgrids, there are better ways for grids to work called smart grids. These are designed to limit waste from big transmission lines and to move energy around in the most efficient way possible. The old grids had a fairly simple task: to distribute energy from a power plant outward to the people who used it. This new technology is important for managing grids where energy is flowing in multiple directions from household producers, and large-scale producers, to storage, to those who use it, and back again.

PG&E is regulated, but as is often the case with public regulation of monopolistic private companies, the agencies doing the regulation have been too close to the company and the regulations have been too weak. In recent years activists have pushed to take the electricity buying power away from PG&E. All around the state and in many other places in the country, energy democracy, where communities control electricity and manage it for the public good, is growing.[5]

In 2002 advocates got a bill passed in the California legislature that allowed local public entities, like towns or counties, to create their own power companies, responsible for buying electricity. In San Mateo County, where I live, the switch was cheaper for everyone from the beginning, and from the beginning the electricity that the new municipal electric company, Peninsula Clean Energy (PCE), provided was cheaper and had a higher percentage of renewables than our old electricity. And we had a choice, for a little bit more money, to get 100 percent carbon-free electricity. Most people I meet have no idea this even happened and that their electric energy is now quite clean. Many of these kinds of systemic changes happen under the radar of popular consciousness. PCE is also investing its profits in developing equitable and sustainable transportation systems in our county.

Huge reductions in greenhouse gas are possible simply by ending the waste from old generation energy distribution systems. And a rapid move to solar and wind is being achieved by taking the fossil fuel companies out of the energy equation, policies that require higher levels of renewables in our electric grids, and leveraging the resources of local rooftop solar and wind energy to meet our energy needs. There is no reason for these systems to be run by private for-profit entities.

Some Significant Transformations in the Energy Sector

- "In the last decade the share of wind and solar in the energy sector doubled every five years. If doubling continues at this pace, fossil fuels will exit this sector by 2050."[6]
- "50% of all new energy capacity in India came from solar in 2018."[7]
- Five African countries—Egypt, South Africa, Kenya, Namibia, and Ghana—installed 1,067 megawatts of new solar electricity in 2018.[8]
- "In 2018 renewable energy accounted for almost two thirds of new installations for electricity generation. Solar and wind energy accounted for 84% of this growth. The world has reached a tipping point in price and performance for wind and solar power. In many places wind and solar are cheaper than fossil fuel alternatives."[9]
- "Emissions have peaked and are now decreasing in 49 countries (accounting for 36% of global greenhouse gas emissions) In 18 developed countries (representing 28% of global emissions), CO_2 emissions declined at an average rate of 2.4% per year between 2005 and 2015."[10]
- "Portugal produces 50% of its yearly electricity from renewables."[11]
- "In 2018 Portugal's CO_2 emissions fell 9% and Ireland's dropped 7%, while their economies grew."[12]

Transportation

One-fourth of global emissions come from transportation. "Urban transport is the single greatest source and growing."[13] If the poor countries of the world follow the path of the US, in having car-based societies, we will not be able to build a sustainable world. Electric cars are a good transitional technology to get us to stop burning gas while our cities develop functioning alternatives to individually owned, single-occupancy vehicles. But cities can't function in sustainable ways with too many single-occupancy vehicles.

The real path to sustainability is to invest in public transportation. In cities with functioning public transportation systems everyone is able to travel for a small price, so those systems increase equity. A city with well-connected and free or inexpensive systems of electrified public transportation will have much less air pollution and much less traffic. It will also have more of a sense of conviviality, where positive human interaction is fostered, as people connect with each other in their daily interactions. Such a city will also have more land for public use, as less land is set aside for parking.

People who live in places with ineffective systems of public transportation, and with huge investments that support automobile use, often can't imagine what it is like to have an attractive system of public transportation. This makes it hard to build public support to invest in public transportation. Buses are often associated with slowness, inefficiency, and the marginalized communities that rely on them. But bus rapid transit (BRT) is one of the most cost-effective ways to quickly develop a functional, equitable, and efficient transportation system. With enough investment in high quality buses, and if buses have priority at traffic signals, systems for payment that are at stations to keep the bus moving, and express lanes, buses are likely to be the most important way that transportation becomes sustainable.

The city of Curitiba, Brazil, built the first and most impressive system of BRT in the world. The project was spearheaded by Mayor Jaime Lerner, who was trained as an architect. Arguing against expensive systems of subways that were designed to serve the wealthy, Lerner famously said, "If you want creativity, cut one zero from the budget. If you want sustainability cut two zeros." He moved the discussion in his city from the idea of expensive trains and polluting cars, to making buses into something that really worked.

Los Angeles has also invested heavily in BRT. Its BRT system costs a fraction of what subways cost and it was much faster to build. As of this writing transit advocates are proposing a reginal express bus system that runs on the freeways of the San Francisco Bay Area. Maps of the proposed system look like subway maps. And the goal is to have people in that system move faster than cars in the current traffic, and eventually, to lead to an end to traffic.

Some Significant Developments in Transportation

- Over two million people a day ride Curitiba's system, and more than two hundred cities in the world now have bus rapid transit.[14]
- In 2018 the country of Luxemburg made all public transportation free.[15]
- "Approximately 62% of Copenhagen's population cycles to work or school."[16]
- "Between 2017 and 2019 eleven economies announced bans on new fossil fuel vehicle sales, or the intention to introduce future bans, including France, UK, China, India, the Netherlands, and Ireland."[17]
- "The "Stay on the Ground" campaign in Sweden, urges people to take a flight-free year. This, and a recently introduced flight tax, may already have had some impact. In the first quarter of 2019 Swedish airports saw a 4.5% decline in passenger numbers compared to the first quarter of 2018, and demand for international train travel is growing simultaneously."[18]

Buildings

Leadership in Energy and Environmental Design (LEED) is an international standard that was developed to rate the greenhouse gas impact of buildings. It is used to rate commercial, institutional, as well as residential buildings. It has standards for new buildings and for retrofitting. LEED includes the greenhouse gas impacts of the materials used to build and run a building, as well as issues of sustainability and equity in how the building is used. For example, the LEED standard includes points for the energy used in the operation of the building. And buildings get points for having gender-neutral bathrooms.

Many buildings are now being built that are net zero, meaning that if they use energy, they have solar panels, which in the course of the year generate as much energy as the building uses. In total they produce no greenhouse gases. The building may have solar panels, good insulation, a system for recycling some of its water, a green roof, good design for where to put windows, and many other design elements. It may have been built from recycled materials.

I teach in a net zero building, and it is a joy to be in. It has a lot of natural light and good temperature regulation. Part of the passive heating and cooling system of the building is a beautiful fountain outside that cycles water through the building. Studies have shown that students learn better in green buildings, which tend to have good ventilation and good natural light.[19] Our building also includes showers to incentivize biking for those who want to shower and change before starting work. These buildings do not cost more to build than standard ones and they cost much less to run.

The Empire State Building in New York City is an impressive example of retrofitting an existing building. In 2008 the building underwent a massive energy efficiency renovation. Heating, cooling, lighting, and window insulation were all redone. The retrofits to the building cost $20 million, but they save $4 million every year. One study found that in the US $279 billion invested in retrofitting homes and institutional and commercial buildings would save $1 trillion in energy. And the work would yield 3.3 million jobs. Doing this would cut US emissions by 10 percent.[20]

Some Significant Transformations in Building Energy Use

- In Vancouver, Canada, "emissions per square meter in new buildings fell by 45% between 2007 and 2017."[21]
- "New York City requires buildings over 25,000 square feet to reduce their emissions by 40% by 2030 and 80% by 2050."[22]
- "Mjøsa Tower [an eighteen-story building in Brumunddal, Norway], the world's tallest wooden building, proves that tall structures can be built using wood. Using timber for buildings can reduce emissions from material production by up to 85%."[23]

Consumer Products

For a long time, when cities, states, and nations calculated their greenhouse gas impacts, they left out the impacts of production and disposal of consumer products bought and discarded in their area. Because so much manufacturing is done in the Global South and consumed in the Global North, this led to a distorted understanding of the greenhouse gas impacts of different populations. Analyses are starting to measure per capita emissions in ways that include what is consumed in a country rather than just what is produced. When analyzed this way, the per capita emissions in the US are twenty-two metric tons of CO_2 equivalent. In the poorest countries of the world they are 1 percent of that.[24] The old way of measuring national impacts made that difference less extreme.

In the old mantra "reduce, reuse, recycle" the most important, and most neglected, part is reducing. Some critical reducing happens with consumer choices, like buying less unnecessary junk. In recent years there has been a rising critique of fast fashion, where we all feel social pressure to buy clothing that is on trend, and which doesn't last very long.[25] There are cultural things we can do to step off that hedonic treadmill, whereby we seek pleasure in consumption, and when we feel sad or insecure again, we go out shopping again. We can buy things that are well made and will last long, shop in vintage stores, and do clothing swaps.

In addition to moving to a society that is not based on overconsumption, two of the most important changes that need to happen in the production and disposal of consumer products are a move away from planned obsolescence and the development of cradle-to-cradle design. Right now, in the US, the incentives are in favor of producing goods that will be discarded and need to be bought again soon, that is planned obsolescence. And in a capitalist economy, it is in the interest of manufacturers to produce things as cheaply as possible, to get us to buy as many of them as possible, and to have them become obsolete, broken, or out of fashion quickly so that we buy new things. Unless governments or consumers hold producers to account for the costs of production and disposal of their products, there are no incentives to produce things in ways that make them have a low environmental impact, to last long, and to leave us satisfied that we don't need more.

Most of us think about waste only when contemplating where to put something we are done with, but far more important than whether the item ends up in the right bin are the energy and materials that went into its production. People are beginning to analyze the chains of commodities that go into making a thing and analyzing the climate impacts of each of those elements.[26]

A recent study showed that all around the world appliances are having shorter life spans. Some of that may be due to planned obsolescence, and some of it is due to consumers preferring to have the latest thing. If people need to buy a new washing machine every five years as opposed to every thirty years, we are wasting huge amounts of energy and resources.[27] Government can set policies that take away the incentives to produce things for short-term use. They can require manufacturers to put an estimate of longevity on the label, and they can make the companies pay for the disposal costs of the item, or take the thing back and be responsible for its disposal.

These policies are part of what is called sustainable material management (SMM). The US Environmental Protection Agency defines SMM as

> a systemic approach to using and reusing materials more productively over their entire life cycles. It represents a change in how our society thinks about the use of natural resources and environmental protection. By examining how materials are used throughout their life cycle, an SMM approach seeks to:
> * Use materials in the most productive way with an emphasis on using less.
> * Reduce toxic chemicals and environmental impacts throughout the material life cycle.
> * Assure we have sufficient resources to meet today's needs and those of the future.

Cradle-to-cradle design is a related concept, which encourages manufacturing in ways that create a circular economy, where products are created in ways that account for how long they last, and hold manufacturers to account for the waste they generate from manufacturing and from disposal.[28] If manufacturers are held responsible for the cradle-to-cradle impacts of their production, our systems would include much less waste.

A capitalist economy working on its own will not favor cradle to cradle manufacturing and sustainable materials management. And it is too hard for consumers to know the environmental impacts of the products they buy to rely on the pressures of consumer choice to solve this problem. To get manufacturers to produce using cradle-to-cradle and SMM practices, governments need to enact regulations that hold manufacturers accountable to these higher standards. They can require smartphone manufacturers to recycle rare earth metals, they can put a tax on items that don't last as long as they should, and they can outlaw unsustainable manufacturing practices. Countries can outlaw the importation of products that don't follow these higher standards.

Some Significant Transformations in Consumer Products

- "The US resale market has grown 21 times faster than the overall retail market over the past three years. It is projected to grow to nearly 1.5 times the size of fast fashion. Resale can be key to a circular fashion industry."[29]
- In 2020 China passed a new set of regulations limiting the use of plastics. "Plastic bags will be banned across all cities and towns in 2022. . . . The restaurant industry must reduce the use of single-use plastic items by 30%. Hotels have been told that they must not offer free single-use plastic items by 2025."[30]
- "France has pledged to use only recycled plastic by 2025."[31]
- "Unilever has committed to 100% reusable, recyclable or compostable plastic packaging by 2025."[32]
- "By July 2019, 90 companies, representing 12.5% of the global fashion market had signed the 2020 Circular Fashion System Commitment to reduce waste."[33]

Agriculture and Food

According to the IPCC 21–37 percent of global emissions come from our food systems.[34] That includes deforestation for livestock, growing food for people and for livestock, producing food, and transporting it. Livestock is a major contributor to deforestation and contributes to 15 percent of total global emissions. The most important part of moving to sustainable food system is encouraging people to eat less unstainable meat. I say unsustainable meat, because it turns out that chicken can have a climate impact as low as many plant-based proteins, when raised right. And different types of meat have radically different climate impacts, with beef being among the worst.[35] Another aspect of sustainability and food is the problem of how wasteful our current systems are. About one-third of all food produced in the world is wasted either in production, sales, or at home.[36] Food waste is linked to 8 percent of total global emissions.[37]

According to Michael Clark, in a study of the relationship between climate-friendly diets and health, "Continuing to eat the way we do threatens societies, through chronic ill health and degradation of Earth's climate, ecosystems and water resources." And while some meat is sustainably raised, "How and where a food is produced affects its environmental impact, but to a much smaller extent than food choice."[38] Globally, there is an inverse relationship between countries with higher levels of meat consumption having higher levels of heart disease, diabetes, and obesity.[39]

Much of the work to shift to climate friendly diets can be personal and cultural, but there needs to be systemic work to speed the shift. Fano and Herrero argue that people as individuals can shift to a more plant-based diet, but "for the world to make this shift, we need governments and the food industry to make it easier. We need investment in public health information and the implementation of policies that promote healthy eating that is affordable, safe, convenient and most of all, tasty."[40]

Industrial agriculture is one of the most destructive aspects of our current world economic system. It requires massive fossil-fuel-based chemical inputs and is causing massive levels of deforestation. It ends up depleting soil, and leading to food with huge carbon footprints. It is more profitable, when working at an industrial scale, to produce food that is high in sugar, salt, and fat, and it is easy to hook people on those foods. The industrial food system is as unhealthy for our individual health as it is unsustainable for the planet.

Our food system needs to be transformed to develop regenerative practices that help develop healthy soil, which actually sequesters, or takes in, carbon. Modern industrial style agriculture is responsible for the degradation of soil.

Sustainable farming practices can actually turn food production into carbon sinks. Those kinds of farms are much better for the people who work on them than the ones which use pesticides and high intensity industrial practices.

It is interesting to note that most of the world produces food in ways having nothing to do with the industrial food system. Mary Robinson argues that "nearly 70 percent of the food consumed around the world is produced by millions of smallholder and subsistence farmers across Asia and Africa—the vast majority women."[41] Those small-scale food systems can be made more sustainable and less labor intensive, with small investments in education, better water systems, and increased access to credit.

Reducing the environmental and social devastation caused by the industrial food system requires challenging the power of entrenched agribusiness interests. In the US, and in many other countries, the meat and dairy industries are heavily subsidized. One important step in reducing emissions is to put an end to those subsidies, and invest in retraining and alternative job creation for people who have been working in those industries. Changing the political support for industrial agriculture, having governments outlaw the use of dangerous pesticides and fertilizers, and investing in the development of sustainable agriculture are crucial steps in the transition to a sustainable food system. Working to promote the health and taste benefits of a low-carbon diet is also important.

Some Significant Transformations Happening in Diet, Food Waste, and Food Production

- "The global vegan food market size is expected to reach US$24.06 billion by 2025 with a yearly growth rate of nearly 10%."[42]
- "Around 60% of Americans report they are cutting back on meat-based products and of these 77% hope this to be a permanent shift in diet. A recent report found that US

beef consumption fell by 19% between 2005 and 2014. In the UK, a recent survey shows that over a quarter (28%) of meat eaters have reduced meat consumption and a further one in seven (14%) adults aim to do so in the future."

- "China has set a target of halving meat consumption by 50% by 2030. This could reduce global agricultural emissions by 12%."[43]

- "In 2016, France outlawed food waste from supermarkets. Italy and Germany have since implemented similar bans."[44]

Planting Trees and Other Carbon Sinks

The sections above focused on ways to decrease the greenhouse gas emissions from major sectors of the economy. This section looks at the other side of the equation: how to pull carbon out of the atmosphere. Most of us have learned by now that plants are the world's lungs. They take in CO_2 and breathe out oxygen. In order to keep those lungs healthy, we need to protect forests, plant trees, and develop other carbon sinks. The book *Drawdown* claims that as of 2016, fifteen billion trees are cut down in the world each year and "Carbon emissions from deforestation and associated land use change are estimated to be 10–15 percent of the world's total."[45]

One of the most urgent problems in this area is the continued deforestation for agriculture, to grow feed for cattle, and for cattle grazing. As we saw in the previous section, lowering global demand for meat is an important part of stopping deforestation, as is strengthening governance in the countries where this is happening. Wealthier countries need to provide funding so that countries in the Global South can stop illegal logging. Generally, when indigenous people control land it is kept in better shape than when it is controlled by national governments. Other small-scale farmers are also often good stewards of the land and their voices also need to be respected.

In 1977 Kenyan activist Wangari Maathai helped found the Green Belt Movement. Her work also led to the planting of hundreds of millions of trees. Trees do not just pull CO_2 out of the atmosphere at an amazing rate. They also restore local ecosystems and thus also lead to more access to water and food products from the local land, while also providing habitat for local plants and animals, protecting against heat in summer, and making places more beautiful and livable. In the Global North many cities also have ambitious programs for planting trees, with many of the same benefits.[46]

As the planet warms and fires become more common, we need to invest in other ways to store carbon, including restoring wetlands and peat bogs, which sequester greenhouse gases. There are efforts all around the world to sequester carbon in soil. The city of San Francisco is embarking on a project to take its municipal compost and spread in on open range land. San Francisco has compost bins for households, businesses, and public institutions. That policy alone diverts those material from landfills, where they otherwise would produce huge quantities of methane. The new project takes that compost and spreads it on rangeland, where it decomposes in ways that enrich the soil and integrate very significant amounts of CO_2 into it, keeping that CO_2 out of the atmosphere.[47]

Some Significant Transformations in Carbon Sinks

- "The Trillion Tree Campaign connects funders with forest conservation ventures, to restore and protect one trillion trees by 2050. A trillion trees planted could potentially capture 25% of all human-made CO_2 emissions."[48]
- "Nature-based solutions, from forest protection, grazing management and fertiliser management, can halve emissions by 2030, while reforestation, biochar and improved agricultural practices have the potential to store up to 9.1 billion tonnes of CO_2e [carbon dioxide equivalent] annually, eventually storing 225 billion tonnes by the end of the century."[49]

- "The Katingan Mentaya Project protects vital peatland habitats in Central Kalimantan, Indonesia. The project has prevented the release of greenhouse gases equivalent to over 30 Mt CO_2."[50]

Investment in Sustainable Poverty Alleviation in the Global South

Along with stopping greenhouse gas emissions and pulling carbon out of the atmosphere, another large-scale sector that is important for a just transition to a sustainable society is working to bring the world's poor out of poverty.[51] This is important because the needs of these people matter. And if they don't come out of poverty in a way that is sustainable, there will be constant pressure to pull them out in unsustainable ways. The fossil fuel industry has used the concept of "energy poverty" to argue that if you care about global poverty you will support fossil fuels projects in the Global South.[52] Focusing on sustainable ways to alleviate poverty leads to better lives for the world's poor and they build political will for a just transition.

Poverty alleviation is important everywhere in the world. Most of us are aware of what is needed to stop poverty in the Global North: taxing the wealthy to pay for social safety nets, programs to increase employment, education to make employment accessible to the poor, and cash for people who are not able to work for a variety of reasons.

The pathway to sustainability in much of the Global South includes those things, but it also needs to deal with the fact that in much of the Global South the infrastructure is not present to access the technologies needed for a comfortable life. Over a billion people in the world have no electricity, and 2.5 billion cook by burning animal dung or wood. These people spend huge amounts of time collecting fuel. Giving them access to solar panels to light their homes and sustainable cookstoves will lead to tremendous difference in their lives. And it will cut carbon emissions. When people chop down forests for cooking fuels there are fewer carbon sinks and less healthy habitat in the forest.

The wealthy countries of the world have emitted 79 percent of greenhouse gases that are in the atmosphere right now, and they continue to emit a disproportionate share.[53] The majority of people, though by no means all, in the wealthier countries have relatively comfortable lifestyles, based on having electricity, roads, and well-constructed homes. It is crucial that the countries of the Global North invest significant sums of money into helping the countries of the Global South to move people out of poverty and into sustainable lifestyles. In many areas of life, they can skip over the dirty resource intensive model of "development' that was taken by the Global North. They don't need phone lines installed. They have generally gone straight to cell phones. And they don't need large-scale wasteful energy grids or fossil-fuel infrastructure. They can go straight to solar power and smart microgrids for electricity. Cities can be developed with public transportation at the core of their planning.

What many of these countries don't have, which they need, is the capital to invest in these projects. For decades transnational organizations such as the International Monetary Fund and World Bank have pushed them to develop projects, such as roads, dams, and fossil-fuel infrastructure, that were good for global capital. And in exchange for those loans they would be forced to cut spending on health care and education. Those transnational organizations have been vectors of capitalist expropriation and high carbon development. In recent years there have been attempts to reform them and get them to provide capital for sustainable projects.

At the international UNFCC climate summit in Copenhagen in 2009, the world's countries established a Green Climate Fund. The wealthier countries of the world have committed over $5 billion to invest in projects that help bring people out of poverty with projects that reduce greenhouse gas emissions and also invest in resilience to climate-change-created dangers. The projects include things like investing in resources and education for women in Ghana to develop climate resilient approaches to agriculture, and developing a water system in the Marshall Islands that is resilient to sea water intrusion as the sea level rises. Many studies have found that educating girls is crucial for bringing a society out of poverty.[54]

Making the transition to a sustainable world will require much larger investments from the wealthy countries to help other countries move out of poverty and into lives that are sustainable in terms of human well-being as well as in terms of greenhouse gas impacts.

Some Significant Transformations in a Just Transition in the Global South

- The Global Alliance for Clean Cookstoves works with people in the Global South to design stoves that work for local conditions. As of 2015 28 million households had gotten access to clean, locally appropriate, cookstoves.[55]
- The project Lighting Africa provides finance to get solar power to people living in rural Africa. As of March 2020, they have helped 32.3 million people access solar power off the grid.[56]
- As of 2017 international organizations spent $13 billion to educate girls.[57]

Resilience

The final big bucket we need to look at for the transition to a sustainable world is resilience. The climate crisis is with us already and some of its impacts, such as sea level rise, are irreversible. Because of this, along with cutting greenhouse gas emissions, pulling CO_2 out of the atmosphere, and investing in poverty alleviation, we also need to develop resilience, so that we can weather the impacts that will continue to get worse. Along with the mitigation that comes from reducing greenhouse gases, we need to engage in adaptation.

Forty percent of the world's population lives within 150 kilometers of a coast.[58] And most of the world's largest cities were built along coasts. There are many things being done to develop resilience in coastal cities. In 2012 after Superstorm Sandy devastated whole communities in New York, the city intensified its planning for storm surges. It is developing barriers to keep storm water out of the subway. It is also planning to create parks on landfill outside the existing edges of the city that have berms to protect the existing parts of the city from storm surges. These newly created parks make the city more enjoyable in the short term, while making it more resilient in times of crisis. Copenhagen is creating a large below-ground park that can catch huge quantities of water when there are storms and be a public amenity the rest of the time.

In much of the world we need to protect coastal wetlands, which act as sponges for storm surges. In the Global South, developing resilience often involves planting mangroves and other similar plants that buffer storm surges, sequester carbon, and provide habitat for plants and animals.

Some Significant Transformations in Developing Resilience

- "Copenhagen's new Enghaveparken will have spaces that can host sporting events during dry weather and fill with water during heavy rains."[59]
- The state of Louisiana is investing in restoring the wetlands that had been destroyed in the years leading up to Hurricane Katrina. That loss was one of the major reasons that storm caused so much devastation.[60]
- The State of Maryland has a permitting program to encourage projects that develop a living shoreline, which absorbs storm surges, rather than an armored one, which just transfers the force of the storm to another place.[61]
- The New York City subway system has installed fluxgates that can be operated by one person and which protect the system from flooding.[62]

- The Institute for Sustainable Infrastructure has developed a rating system, Envision, which, like the LEED program for building design, can be used to analyze the resilience of new infrastructure projects.[63]

Conclusion

Huge things to make human society sustainable are happening all around the world, and the learnings from those projects are being shared at lightning speed. The technology needed to completely eliminate the use of fossil fuels exists, and new clean energy is cheaper than using fossil fuels. We need to stop cutting down forests, plant trees, restore wetlands, put carbon into the soil, and pull carbon out of the atmosphere. We have figured out ways to feed everyone in the world a healthy diet in ways that are sustainable. Techniques have been developed to make our cities and coastlines more resilient to the storms that are coming and to make agriculture resilient to drought. And much of that transition is happening at a rapid rate in every sector of most countries of the world. Millions of people are already hard at work on those transitions.

To make these changes at the speed necessary, governments will need to enact regulations and direct funds to make them happen. That will require a shift from a hands-off approach to government and it will require that fossil fuel companies are not allowed to slow down progress. That is where climate justice activists are most needed.

The hard question for our future is not what to do, but how to get our systems to do those things quickly enough to avoid the worst outcomes. So many of the innovations outlined in this chapter will make our lives better as they also make our world more just and environmentally sustainable. There is no need for huge sacrifice and painful choices to have a sustainable future. Climate justice activists need to shine a light on the forces that are getting in the way of rapid adoption of these sustainable practices, fight as hard as we can to get the fossil fuel industry out of our social decision-making processes, and get massive investment in the transition.

CHAPTER 5

Advice for Action

Making a just transition to a sustainable society will require an unprecedented mobilization of social power. And different people will find their place in that work at many different levels, and with many different kinds of action. Chapter 8 explores the wide variety of ways people can engage and lays out the different things to consider for yourself when you are deciding what kind of actions to take. This chapter looks in a general way at the kinds of impacts that climate action work has in three main areas: personal lifestyle changes, organizing for cultural change, and organizing for policy change. Each has its own benefits and limitations, and they are all interrelated in complex ways. The goal of this chapter is to help you understand in general terms how social change happens, and how to be effective in the actions you take.

Personal Lifestyle Changes

For years environmentalists have told us to turn off the lights, carry a water bottle, and reduce, reuse, recycle. There are many tools available on the internet for tracking a person's carbon footprint. And to the extent that one can do it, it is a good idea to make these changes. Planting one tree pulls carbon out of the air. Changing to a more plant-based diet is important. Flying less, driving less, and consuming fewer wasteful products are also important. Less carbon is less carbon.

These are all good things, and each of our individual actions really do make a difference. But it is crucial to understand how those individual choices make a difference, and how to get them to be as effective as possible. And it is important to understand the limitations of an individualist approach. For some people, these individual choices can be a first step in taking more impactful action for the climate. Sometimes a person becomes aware of the severity of the ecological crisis we face, and that awareness motivates them to make different personal lifestyle choices, and that can lead to engagement with collective forms of action. But we need to understand how these individual choices can add up to more than the sum of their parts and what kinds of changes we need to push for that go beyond them. We also need to be mindful that a focus on individual consumption can sometimes be counterproductive.

The large powers profiting from the current state of affairs have an interest in having us only do small personal things, and to not work for systemic change. The idea of individuals tracking their own carbon footprints became popular as a result of a 2005 campaign promoted by the fossil fuel company BP.[1] The iconic antilitter campaign in the US in the 1970s was paid for by beverage manufacturers who used that campaign to take attention off the waste generated by manufacturing and onto the problem of putting the disposable container into a trash can.[2]

For some people, who have the money and time, living green is a badge of honor. Many people choose to insulate their homes, buy solar panels, drive electric cars, grow vegetables, and compost in their gardens. Those are all great choices for people who can make them. But on a broader level, those environmental lifestyle choices can also backfire, if we are not careful how we talk about them. If expensive and time-consuming actions are what it takes to be an environmentalist, and a person can't install solar panels, can't afford a low emission car, or doesn't have the time to grow vegetables and compost, they can feel like environmentalism is simply not for them. If we are not careful, a personal lifestyle focus can trigger increased defensiveness, as it also triggers feelings of guilt and shame.

Many of the people living a green consumerist lifestyle can have very high carbon footprints. The place in the US where it is easiest to live with the lowest carbon footprint is New York City. Living in small homes, and using public transportation are great for a person's carbon footprint. So is being poor and not buying very much stuff. Many wealthy environmentalists live in large houses, which require a lot of energy to heat and cool, they buy a lot of new things, and they often have the privilege to fly a lot for pleasure, often on eco-adventures.

At its worst, this individualized approach can lead to dysfunctional forms of green consumerism, where for some people using green products makes them feel so virtuous that it can lead them to justify flying from Europe to Thailand for an eco-vacation. Per Espen Stoknes argues that "pushing personal lifestyle changes can make us complacent and less vocal for change on the political and social level.... Climate change isn't an individual, technical, or environmental problem. It's a cultural challenge, with solutions at the organizational social level."[3]

An individual focus can also be counterproductive when it leads to people feeling guilty when they think about the environment. Most people rely on carbon-based fuel to heat or cool their homes, most use it to get to work, and some rely on it when they fly to visit their families. And we use polluting fossil-fuel-based plastics when we buy the products that we feel we need to live lives that are in sync with the people around us. Most people throw their food scraps into the garbage because there are no easy alternatives. And for many—because every day they choose how much to drive, to turn the heater on or not, what to buy—when they think about the climate crisis, they feel a sense of shame. When faced with the ways our lives are deeply embedded in the matrix of fossil fuel use, many people simply turn away and say that isn't their fault, rather than face that shame.

It is important that we are deeply mindful of the psychological traps that prevent people from taking strong action for the climate. When our society frames climate action as an individual choice to buy different products, that can lead others to see the crisis in individualistic terms, and that can provoke unhelpful forms of guilt. And that sense of guilt can be counterproductive to action. The focus on personal lifestyle choices in the environmental movement has led to an impoverishment of the ways we imagine what we need to do to make a difference.

Impactful climate action, at the scale needed to turn around the climate crisis, is often something quite different from personal lifestyle changes. And it requires us to do things that are very different from what society tells us we are supposed to do. To get to the large-scale rapid forms of action we need, we need loosen the grip of individualism and guilt on the climate action imagination. We need to see our actions as part of a larger movement to transform society such that fossil fuels are not burned and forests are rebuilt rather than destroyed. Our individual actions are important if they shift the culture toward widespread use of good individual practices, or if they lead to pressure to change social practices such that individual environmental choices are easier to make.

And some things are more amenable to this sort of individual approach than others. For example, most forms of meat are truly terrible for the environment. People who eat less meat can spread the idea that there are lots of other great things to eat, and they can be healthier. But how one talks about it makes a big difference. We can talk about it as a great delicious choice, rather than as a painful sacrifice, or as the thing you must do to be a good global citizen.

Taking the bus when you have the option to drive is a good thing to do, and it might inspire others who could drive to choose the bus. This can be a personal lifestyle choice. But it will be so much more effective if you also get involved with advocacy and push for better bus service.

At De Anza College, where I work, in 2013 students advocated for a bus pass system. Everyone paid $15 annually and got a bus pass that was worth $700 per year. The system almost immediately instituted an express bus because the number of students going on that route increased dramatically. The work the students did to get the bus pass was much more significant than having each of them choose to ride a bad bus system. The work took them a few years, but it was also an engaging process, and the people doing it forged powerful relationships, and still years later they remember their work with pride.

In the town of Pacifica, California, where I live, the city picks up compost. That makes it very easy to compost food and garden waste. The work that helped to make composting easy for everyone in our town was far more important than me individually putting a banana peel into my backpack and bringing it home because where I am doesn't have compost. There is nothing wrong with putting banana peels into one's backpack, but an overemphasis on the virtue of that action, can distract from the larger work of making the social changes that make good choices easy for larger groups of people.

One of the most pernicious myths that drives a focus on personal lifestyle choices is the myth that markets solve social problems. The understanding of society that people are taught in most Introductory Economic classes says that we are all individuals, connected by the markets that magically coordinate the dance of production and consumption. In fact, human beings are tied together by a mass of connections into deeply interwoven fabric we call society. That society is made up of a political system with all sorts of rules and regulations, and it is made up of systems of meaning through which we all come to understand our lives.

In a society saturated with procapitalist ways of understanding the world, people tend to be pessimistic about working through government and other institutions to get changes that society needs. And they are constantly told that they should pursue their own pleasures and the market will solve social problems and get all of our needs met. And so many people believe that they should act in good ways as an individual and then the market will adjust and provide good products. The problem is that market signals we send as individuals are actually very small, and usually there are structural barriers that need to be addressed, and that require intentional policy to change.

The move from gas to electric cars is a good example of the weakness of market signals to get a major transformation, and of the role that individual choice plays as part of the equation. Electric cars need a charging infrastructure, and manufacturers wouldn't move to the new technology if they thought only a few people were going to buy their cars.

In California, the state passed laws that require any manufacturer selling cars in the state to have a certain percentage of cars be electric. This forced manufacturers to experiment with developing the technology. Many manufacturers produced a few cars to be in compliance, without really committing to make the change in a big way. A few existing companies, like Nissan and Chevrolet, invested in what they thought was a coming technology and developed mass-produced electric cars. At the same time, the state invested in chargers. Finally, because at the beginning the development many technologies are quite expensive, these cars needed to be subsidized to make them affordable.

As I write this the federal government gives a $7,500 tax break for an all-electric car, the state of California gives, $1,500. My local community-owned electric company, which is investing what would have been profits into transformational practices, worked with car dealerships to knock another $5,000 off the price. Altogether these policy choices made it such that for people who buy cars in my area, buying an electric car is much cheaper than a comparable internal combustion engine car.

The individual consumer is still a part of the electric car equation. In the beginning, it took a bold green consumer to buy a product that others thought might be a waste of money and would be inconvenient. Those early green consumers played an important role. But that role of buying the car is only a small part of a much larger equation, which required huge amounts of policy work, and not a small amount of cultural work as well, to get the word out that it isn't inconvenient to drive an electric car, to make it clear that it isn't just for the wealthy, to make it cool, rather than stupid, to buy one.

Market mechanisms do work to get businesses to fill niches where they can sell things for a profit. But markets exist in the context of governmental regulations, social norms, and the physical limits of the planet. There has never been a free and unconstrained market. Much work done for environmental transformation is about constraining and shaping markets. So, it is important for people doing social change work to challenge the deeply entrenched feelings people have that their only power is as consumers. We need to help people see that we are not just passive consumers and recipients of what life has to give us but are all citizens of the world, who have a stake in it, and whose actions actually create and recreate the world of possibilities we all face.

Individual actions can be good. And they can often be cheaper that the alternative. Gas-guzzling cars and trucks are generally more expensive than fuel-efficient cars. Small homes are generally less expensive than large ones. And buying less fast fashion is cheaper that trying to keep up with trends. I have saved a lot of money in recent years by flying less. Those kinds of individual choices can begin to make a significant difference if the idea behind them spreads. If my not flying is seen as a good choice by others I can act as a role model. But if I am too self-righteous about it, I may lead others to find environmentalism insufferable and give them another excuse to turn off completely to messages about the climate crisis.

Organizing to Shift the Culture
The fossil-fuel-funded climate denial movement did amazingly effective work shifting the culture around the climate

crisis. The public relations firms who fought the move to regulate tobacco products learned that they didn't need to convince all scientists that tobacco didn't cause cancer. They just needed to do a lot of publicity around a few who would carry their message. That was enough to opening a gap in the sense that there was a scientific consensus.

The fossil fuel companies hired those same public relations firms, and spent billions propagating the view that there was something less than a consensus around the science of climate change. After their work, climate change came to be seen as a controversial topic that would be best to avoid in polite conversation. Weather reporters often talk about unusual hurricanes or extreme droughts without saying anything about the climate crisis.

This left a strong imprint on US culture where, for many years, it felt almost impolite to talk about climate change. And the environmentalists didn't help who talked about it as a crisis that was everywhere and nowhere, and that would require us to each make sacrifices as individuals. As we saw in chapter 3, there are all sorts of psychological mechanisms that unfortunately make it difficult to get climate messaging across.

As I write this the most exciting thing happening in the climate action movement is the entry of youth into the conversation. They have inserted a sense of urgency that has resonated with the culture at large. And the focus on the Green New Deal, helps us to see that it is possible to do the things needed to be sustainable and that for most of us, our lives will be better once we do those things. The cultural work we need to do around the climate crisis involves good messaging, and bold actions that shift our collective sense of urgency.

There are many things we can do to shift a culture. Culture shift is one of the main purposes of a demonstration. A lot of people in the streets says that those people think this thing is important, and it forces the news media to discuss it. The slow, long work of talking to people, spreading ideas on social media, and writing letters to the editor are all things an individual can do to shift the culture. Those organizing effective actions think carefully about the messages they are sending. What messages should be on the signs that are given to people who come? Is there something theatrical that can be done that will capture people's imaginations and leave a strong imprint?

Culture change involves work done to change how people think or feel about something. It is related to policy work, because once people think or feel differently, it is easier to enact policies that then institutionalize those changes.

Sometimes getting a culture to change is a result of policies, such as policies over what is taught in schools. In 2013 a group of students came to me and said we needed to require all De Anza students to take a class on sustainability. After many years of hard policy work, we passed a requirement that all students graduating from our college would take at least one class that includes sustainability content.

Another form of culture change work that is related to policy work is passing resolutions. After the world agreed to the Kyoto Protocol but the US would not approve it, the mayor of Seattle, who was doing a lot of sustainability work, developed a resolution for cities to sign onto called the Mayor's Climate Commitment. Hundreds of cities signed on. In some ways you could criticize it as an empty gesture. Unlike the policy work we will look at the next section, resolutions don't bind anyone to do anything in particular and they don't have the force of law. But what they do is to shift the conversation, and so shift the culture in ways that create openings for policy change.

As I write this, cities, counties, and states are passing climate emergency resolutions. Two things happen when your municipality has declared a climate emergency. The first is that the process of arguing and advocating in the public for it shifts consciousness and touches everyone involved with that conversation. Second, it creates an opening for more specific policy action. "How can you vote against the city buying electric vehicles for its fleet, when you have said that there is a climate emergency?" Cultural change can make policy change easier. And of course, the relationship is cyclical. Good policies, such as municipal compost, then make individual actions easier and more impactful, which shift cultural norms about how we deal with garbage.

One important thing to remember when working on cultural change is that people belong to subcultures. They form identities and senses of self in connection with others. It is almost impossible for a liberal environmentalist to change the views of a committed conservative from the position of that environmentalist's worldview. Instead, anyone wanting to work across subcultures needs to think of ways to frame messages that resonate with the people in the other group.

In many conservative states in the US, such as North Dakota, wind and solar installations are going up at a rapid rate. The farmers who put windmills on their land are doing it because someone is offering them money to do so. And if they are part of a conservative cultural group, a barrier to taking that good deal is likely to be the ribbing they will get from their friends about becoming a member of the liberal elite. Many environmental changes can be promoted as money savers to people who do not believe in the reality of climate change. If appliances that use less energy have the amount of money that will be saved over the lifecycle of the appliance written right on the tag, that will make it more attractive to buy. And packaging it as the green product may actually be counterproductive for selling them to people who are embedded in anti-environmentalist subcultures. The federal government's energy rating system is called Energy Star. It is not called Green Choice. And that is for a reason.

And a surprising thing is that, as the world changes around them, people change too. As Per Espen Stoknes argues, we are herd animals, and some of us belong to different herds. You move your political agenda forward by making the world make sense in ways that build support for your view from the people who will enable you to do the practical things you want to do.

The great Italian philosopher Antonio Gramsci argued that an important part of politics is *hegemony*, or a world view that makes the world make sense in ways that enable us to realize our goals. Gramsci's most famous example was the way the Catholic Church in Italy in the early part of the twentieth century helped the economic elite. The church asked people to be meek and expect rewards in an afterlife. This way of understanding their place in the world made the work of those wishing to organize the Italian working class to resist exploitation difficult. The idea that good people are meek became common sense. So, a big part of the organizing on the left had to help people to understand the world in a different way, to create a new commons sense, and there- fore, on a political level, to create a path to what he called a counterhegemonic worldview. Hegemony is the building of power through creating systems of common sense which empower a dominating agenda. Counterhegemony means building systems of meaning that help drive efforts to chal- lenge political domination.[4]

Social movement theorists talk about how alternative ways of doing things often start with a small group of people who seem to be out of step with society.[5] That small group can develop counterhegemonic way of understanding the world. New and different ways of understanding the world often start out as marginal and are ignored or made fun of. Over time the people with that counterhegemonic approach to life engage with the public, take actions, such as demonstrations, that force a larger society to think about their message. Then after a while people have to take sides on the issues, and many come to sympathize with the alternative way of seeing things. That allows for a political opening that can be used to create policy change. Once the new understanding is embedded in policies, it becomes part of the common sense understanding of the world. The process of going through all of these steps can take years and sometimes can move very quickly.

Protest, Nonviolent Direct Action and Civil Disobedience

One of the most impactful ways to jolt a system into a cultural shift is through public forms of protest. Many movements have made significant progress by using what is coming to be called nonviolent direct action (NVDA), or what used to be called civil disobedience. This was one of the most important strategies of the civil rights movement in the US and also the independence movement in India. Activists in the US used it successfully to stop a meeting of the World Trade Organization in Seattle in 1999. The basic idea behind civil disobedience is that activists interrupt a social pattern that they see as illegitimate in order to bring attention and a sense of urgency to their demands. They break the law in order to change the law or to change social patterns.

A crucial part of nonviolent direct action is that one is trying to bring attention to some sort of social wrong to get the rest of society that is watching to feel that your cause is right. One of the most powerful impacts of the lunch counter sit-ins, which were a crucial early part to the US civil rights movement, was that some well-behaved young Black people sat at a counter and ordered lunch, and were met with brutal repression. The contrast between their comportment and the comportment of the police helped build support for changing the legal apartheid system we had in the US at that time.

In the protest in Seattle in 1999 against the World Trade Organization (WTO) the more confrontational actions that actually shut the meetings down were important for the level of attention that was received. And because most people in the US had never heard of the WTO, the drama of the action helped millions of people ask for the first time what the WTO was and wonder whether the protestors were right that the WTO was a corporate power grab. One aspect of the protests was that as a result of some protesters engaging in property destruction media coverage of the protests turned to a focus on the developing conflict between activists and police. At that point activists were in danger of losing control of the narrative and having the press report about that battle rather than about the WTO.

When Extinction Rebellion shut down the city of London in 2019, they helped spread a sense of urgency. Soon Extinction Rebellion chapters were popping up all over the world. Extinction Rebellion showed a lesson that students of social change have known for a long time: disruption can be an amazing form of pressure. Social movement theorists Frances Fox Piven and Richard Cloward showed in their book *Poor People's Movements* that even though most people will tell you that disruption is counterproductive, because it loses you support, actually disruption has been shown to be one of the most effective tools that marginalized people have to make social change happen.[6] When effective, it can force a quick change in a societies' common sense.

Organizing for Policy Change

So much of what determines how societies function are the rules, policies, and laws that are enacted by governments and other institutions. Governments are places where power is highly concentrated and regulated. They are sites where forces with a variety of agendas vie for power. One of my favorites quotes from the world of organizing is "Laws, policy, and rules are power frozen in time." That means that as we struggle to change laws and we engage in challenging and changing power relations in governments, the results of those battles end up frozen in the actual decision that governments or other institutions make, and they become frozen in the ways society moves forward.

Much of the work that needs to be done to get significant action for a just transition to a sustainable society requires changing laws, polices, and rules. And making those changes happen requires hard work, often work that challenges entrenched systems of power. Some of the work takes place in the cultural realm, where we work to shift how people see the world. But it is when those shifts of meaning create openings to actually change the laws, rules, and policies of an institution or government, that we are likely to get the depth of changes needed to make the transition to a sustainable society.

Some of the most impactful work we can do is in the realm of organizing for policy change. And it is an area of work that is not especially familiar or within the comfort zone of many people. For us to make the transition to a sustainable society we will need millions more climate organizers working steadily and deliberately to change the ways our societies function. While activists are people who are willing to put themselves forward and make a statement, being an organizer is an even deeper level of commitment: it requires that we work in thoughtful and strategic ways to achieve our goals to change policies, rules, and laws.

Effective organizing to change policies is based on a complex set of skills and those skills can be learned. Most people learn them on the job as they work in organization, but there are training programs one can attend, toolkits one can use, and books one can read. Investing in one's skills as an organizer can really pay off in terms of how effectively one uses one's time. There are several excellent toolkits available online, including Beautiful Trouble.[7] One of the best books on this is *Tools for Radical Democracy: How to Organize for Power in Your Community.*[8] For anyone serious about organizing it is good to go to an organizer training and learn the art of organizing. The Community Learning Partnership has organizer training programs, mostly at community colleges, all around the US.[9]

The following section introduces some of the major tools one learns in those trainings. These are: power analysis, issue identification, emotionally intelligent leadership, and leadership styles.

Power Analysis

When organizing for policy change, it is important to be strategic. Often activists pick a tool because it is the first thing that comes to their mind. They themselves notice an issue when people do street demonstrations, so street demonstrations must be the tool for social change. One demonstration on its own, however, almost never leads to policy change or even lasting cultural change. Instead, those demonstrations are usually part of a larger multi-faceted campaign. With a campaign, a group of people spend some time thinking about their goals. They analyze the things that need to happen to achieve their goals, and they execute that plan, while constantly revising as new information comes in. There are many tools in the organizer's toolbox, and the most important one is being strategic and thinking about what all the steps are to make the specific change they are looking for. The best tool for engaging in strategic thinking is called *power analysis*.

When doing a power analysis, one decides on a clear achievable objective, analyses the obstacles to achieving that objective, and comes up with a plan for overcoming each of those obstacles. Often a power analysis involves identifying a target, defined as the powerholder whose action needs to be changed to achieve the objective.[10]

For example, imagine you are trying to pass a local law and your city council has five members. Maybe two of the members are inclined in your direction, two are dead set against you, and one is wobbly. You need to figure out what you need to do to get that wobbly person to vote your way. It may be that exposing the financial connections of the two who are against you will move that person, as they don't want to be associated with their actions. It may be that gaining strong support from your two allies and getting them to bring the middle person along is the best approach. Or it may be that that wobbly councilmember is close friends with someone you know, or maybe their election depended on support from a union you can mobilize to put pressure on them.

The idea of a power analysis is to clearly analyze the field of power relations you are confronting and do the deep thinking and planning that puts you in a position to win your achievable goal.

Issue Analysis

The tool you need to do before you do your power analysis to decide on a good objective is called *issue analysis.* An issue is a slightly counterintuitive term for the specific goal of your organizing, like passing a city council resolution, getting an institution to divest from fossil fuels, or getting a state to outlaw fossil fuel extraction. In *Tools for Radical Democracy*, Minieri and Getsos argue that society is full of problems, like too much greenhouse gas, but that problems are too general to organize around. One of the first steps in an organizing campaign is to turn a problem into an issue that is winnable, such as getting your city to move to municipal composting. An issue, then is the specific solution to part of the problem you are looking at. According to Minieri and Getsos, there are six elements to look at when deciding on which issue to organize around:

- *It resonates.* The issue is important to the people who experience the problem it resolves. It makes them animated and angry.
- *It delivers.* Winning on the issue delivers a concrete, positive change that will make people's lives better.
- *It's winnable.* It is possible to win what you want.
- *It has a clear target.* There is one specific person who can give members what they want. This person is the target of the campaign.
- *It builds power.* The issue strengthens your organization and brings in new resources—members, relationships, allies, funds. It unites rather than divides people in the organization.
- *It supports your mission and values.* The right issue is within the parameters of your organization's work.[11]

One important thing to always remember when doing organizing is that a big part of your goal is to build collective power. A campaign should bring people into your organization, develop their leadership, and develop their sense of commitment and togetherness. If an action is too ambitious, it may burn people out and actually end up weakening the organization. Relationships are key to organizing. People need to feel good, feel effective, and feel appreciated. Doing the work needs to make their life better not worse.

Emotionally Intelligent Leadership and Leadership Styles

When we use the word *leadership* most people think of the charismatic person who is the public face of an organization. In fact, there are many elements of leadership and that public leader, the inspiring visionary, is only type. The organizer training group Wellstone Action developed a nice typology of five kinds of leaders to help us value a range of kinds of action that any organization needs to be effective.

Wellstone Action argues that leadership is made up of five major elements: there are process leaders, who attend to the group dynamics; ethical leaders, who attend to whether what is being done is right; task leaders, who are good at doing and delegating the specific jobs that need to be done to accomplish your goals; strategic leaders, who think about how small actions can build to real achievements; and visionary leaders, who inspire people to keep at the work. Most of us are stronger and weaker in a few of these areas. It is important in an organization to make sure that you have a balance of all five aspects of leadership and that they are working in synergy.[12]

Organizations that function well are generally leaderful. That means that they work in ways that develop the capacities and sense of responsibility of many people. And if those leaders are both emotionally intelligent and culturally responsive, a group and be incredibly powerful. Emotionally

intelligent leaders are ones who attend to the social dynamics of a group, are empathetic toward others, and find ways
to bring out the best in everyone around them. Culturally
responsive leaders are aware of the subtle and not-so-subtle
dynamics of interpersonal oppression and work to have their
group function in ways that run counter to those dynamics.
And so, for example, it will elevate the voices of people of
color, young people, and women in an organization. They will
make sure that a group functions in ways that are friendly
and welcoming to queer people.

One important tool for bringing people into alignment
with the goals of your organizing is *public narrative*, a tool
developed by Marshall Ganz.[13] The idea of a public narrative is that when we speak to others about the work we are
doing we should take every opportunity we have to speak
with some personal revelation about what brought us into
the work. People are more trusting of, and inspired to action
by, people who show some vulnerability and seem like real
and passionate people rather than machines trying to achieve
objectives. The public narrative is a simple short speech that
has three elements: the story of me, the story of us, and the
story of now. It begins with why you personally are invested
in the work you do and then invites your listener to be a part
of the circle of concern you are addressing, so it explains why
they should care about what you are working on, and it ends
with a clear and timely call to action. Practicing a public narrative can make you ready to share your story and be more
able to motivate others to action when the opportunities arise.

Conclusion

Impactful work to build a just transition to a sustainable
society will require everyone to shift our personal lifestyles
around what we consume. But the most important way to get
those changes is to shift the policies that shape the choices
people face. Organizing to shift the culture adds power to
those implementing policy changes. And policies change

when people roll up their sleeves and engage in strategic, con-
certed, and committed action to challenge the power structures that keep harmful policies in place.

CHAPTER 6

Choosing the Best Policy Tools

Work in climate policy is crucial to achieving a transition to a just, sustainable society. Individuals can change their buying habits, and movements can pressure systems to change, but in order to achieve the large-scale transformations needed, in the timeframe that is necessary, the rules that govern how society works and how resources are allocated will need to change.

We will need to enact laws on how electricity is generated and distributed, on what kinds of products can be produced, and who is responsible for the environmental consequences of their production and disposal, on how buildings are made, where which kinds of public transportation and housing are built, and to what standards. It is one of the great myths of procapitalist thinking that social decisions can be made by each of us choosing where to spend our money. Social systems change when the rules under which they operate change. And public policy is the name for the initiatives that set those rules.

There are many avenues for work in this area, and there is much to know to understand the different actions organizations are taking to impact policy. This chapter explores the benefits and limits of regulation, elections, carbon taxes, cap and trade, offsets, green development funds, subsidies, lawsuits, divestment, and investment.

Even if, as an activist, you are not interested in being involved with policy work, and you only want to focus on mobilizing people for protest action, that protest is likely to be aimed at getting the political system to shift in some ways. As an activist working to transform society and get it off fossil fuels, it is unavoidable that in some ways you will be dealing, either directly or indirectly, with these big policy tools.

Regulation

So much of what activist work for are regulations. It is such a common demand that it almost gets forgotten about. Regulations are the rules that governments set for what people and businesses are able to do. Regulatory work can take place in small towns, states, countries, or at the international level. It is always good to look for what the venue is where something is likely to be able to be achieved. In the small town of Pacifica where I live, we worked hard to get a progressive city council elected. While they were in power, we were able to stop a major highway expansion that would have led to more individual cars on the road. Then two years later we were defeated, and as of this writing we have a pro-real estate city council. At this time, it doesn't make sense to ask much of our present city council. We will be trying in the next election to take the majority again. And in the meantime, my group, the Pacifica Climate Committee, is shifting its gaze to work at the county level, since that is a venue that at this time is open to ideas that promote a just transition to a sustainable society.

Doing social change work well involves analyzing the hand you have been dealt, doing what you can to change that hand, and developing smart strategies accordingly. This is an imprecise science. One never knows what will work and what won't, but having a plan and developing strategies make it more likely that your work will be rewarded with success.

In California, where I live, right now we have a democratic majority and a moderate level of support for good policy. Groups are working constantly to bring good ideas to sympathetic legislators. Often they write bills and give them to the legislators. Often policies are tried out in small towns before they are passed at the state level. But some things are better started at a larger scale first. For example, a town can't set high vehicle emissions standards.

Doing work on the inside to get policy-level change is slow and painful. Often a bill starts out being something really great, and by the time it is passed it is so watered down by compromises from the legislators who do the bidding of the fossil fuel industry, that the original supporters no longer support it. How much those inside players get from legislators is deeply connected to the work done by outsiders to put pressure on the system, to create a sense of urgency, to expose the financial ties legislators have to the fossil fuel industry, to expose unnecessary compromises.

Elections

One of the limits on work to change regulations are the people at the table writing and voting on those regulations. You can apply community pressure to politicians, and under the right circumstances even politicians who are opposed to your point of view, and committed to the interests of polluting corporations, can be pressured to do good things.

A famous example of this in the US is President Nixon, who campaigned on his hostility to the environmental movement, yet who signed some of the most important environmental laws in US history: the Clean Air Act, the Clean Water Act, and the Endangered Species Act. He did not do that because leaders talked to him and helped him to see the light. Rather, while he was president there was a massive environmental movement on the streets and in the halls of Congress, pushing for these things. As Nixon tried to hold together a country that was coming apart at its seams, he decided it was more in his interest to placate those forces than to fight them.

In 2016, Alexandria Ocasio-Cortez worked on Bernie Sanders's presidential campaign. After that race was over, she was encouraged by people in that movement to run for a congressional seat and, with a lot of grassroots work, won her seat in the Bronx and Queens, New York. Along with an older, more mainstream liberal senator, Edward Markey, she developed the Green New Deal, the first proposal in US history that is adequate to the real challenge of the climate crisis. The movement to have progressives challenge mainstream Democrats in congressional and local elections, gained a lot of momentum with the 2016 presidential campaign of Bernie Sanders and the electoral successes of congressional candidates in 2018. These victories showed that many things are possible right now that seemed unthinkable just a few years ago.

These political shifts do not happen because a charismatic individual has stepped into the limelight. Rather, good candidates emerge out of movements. Movements build campaigns to get them elected, and movements need to stay engaged to ensure that the politicians they support remain accountable to them. Without this relationship to a movement, it is very difficult to mobilize enough power to defeat the moneyed interests. Without movements staying engaged, those interests are always ready to buy off the people who get elected, no matter how good they are as individuals.

Electing progressive politicians is not an easy task. In the US, and in most countries with formal representative democracy, there is an incredible advantage given by money and by connection to entrenched interests. Yet organizations are developing all over the country, and in many countries in the world, to challenge politics as usual and elect people based on grassroots power, which can be mobilized to defeat corporate interests.

Carbon Taxes

While climate justice activists generally oppose carbon taxes, there are many in the climate movement who see a carbon tax as an important policy tool. The idea is that if you put a tax on fossil fuel use, then using fossil fuels will be more expensive, and the market will help reduce their use.

Many advocates of carbon taxes also argue for tax and dividend systems. In a society with extreme inequality, it is wrong to ask the poor to pay more to heat their homes. Carbon tax and dividend systems give money back to households through tax rebates. It is hard to see a carbon tax being popular with the voting public without a dividend. And even with a dividend, one of the things that makes the passage of a carbon tax difficult is that many people don't like taxes and don't trust their governments to distribute the dividend fairly.

In 2018 Canada enacted a carbon tax, which puts a fee on carbon-based fuels, such as gas, oil, and coal. The Canadian tax began at twenty dollars per ton of CO_2 in 2019 and is slated to rise to fifty dollars per ton in 2022. In the Canadian system, 90 percent of the revenue generated from the tax is refunded to families. Because not all of the tax is translated directly into to higher prices, most people actually come out ahead by a few hundred dollars per year. Every carbon tax and dividend system has its own complex mechanisms for figuring out how much people get back. The basic idea is to make the use of fossil fuels relatively more expensive than clean alternatives, without making people worse off. The rebates help build public support for these efforts.

Businesses generally favor market mechanisms, and so many business leaders and probusiness politicians like carbon taxes. In 2008, Exxon-Mobil CEO Rex Tillerson argued that carbon taxes were the best policy for dealing with climate change. For many fossil fuels companies, a carbon tax is a way of staving off regulation, which they find more disruptive to their business model than taxes. Those companies would prefer higher taxes to higher vehicle mileage standards or to directly regulating how much CO_2 companies are allowed to emit.

At present in the US, the Citizens Climate Lobby has a bill proposed in the US government called the Energy Innovation and Carbon Dividend Act. This act would institute a tax and dividend system, similar to the one in Canada. Supporters claim that the act is bipartisan because, as of this writing, it has one Republican senator on board. In order to get that one Republican, they needed to add to the bill the stipulation that the Environmental Protection Agency would not be able to regulate greenhouse gases for the first ten years of the bill's life. Carbon taxes linked with dividend can be helpful policy tools, but when they are used to replace direct and effective regulation, or investment in alternatives, are dangerously counterproductive.

Even when carbon taxes are combined with a dividend, and don't include the elimination of other important policy tools, there are a few things that make a carbon tax a less than ideal strategy. One is that it assumes that markets work to get good social results. It turns out that when gas gets more expensive, most people drive roughly the same amount. Their lives are constructed in ways that make driving necessary. The much more successful ways that societies have gotten people to use less gas have been direct regulations that force car manufacturers to have higher milage standards, or investment in high quality public transportation. The best, most deeply impactful, solutions around transportation, such as the development of real high-functioning public transportation and housing around transportation hubs, come from other forms of policy, not from a better functioning market, which is what a carbon tax offers. The Canadian carbon tax is set up in ways that make it likely to have a positive impact. But it is important that it be seen as one tool among many, not as an alternative to strong regulations and investment in alternatives.

Cap and Trade

If a carbon tax has been controversial in the climate movement, cap and trade has led practically to all-out war between parts of the movement. A cap-and-trade system caps the level of pollution a company is allowed to emit and allows it to sell its ability to pollute if it reduces emission below that allowed level. On the face of it there is much to dislike in cap and trade. How can we give companies a right to pollute? Yet current regulations actually do just that—they set the allowable level of pollution.

When done right, a limited cap-and-trade program can incentivize companies to lower their emissions more quickly than they might otherwise do. The idea behind cap and trade is that a government will set a cap on how much emissions a polluter, perhaps an energy provider or a manufacturing company, can produce. It then allows the company to trade away the amount of pollution that is left over as its allowable emissions at the end of the year. By reducing its emissions and trading that potential to another entity that hasn't reduced its own emissions, the trading will incentivize that first company to keep reducing even further.

Imagine a company that was considering investing in new green technology. Imagine that company was allowed, by current regulation, to produce five hundred metric tons of CO_2 per year. Imagine that the new technology was expensive, but if the company bought it, it would be able to reduce emissions by one thousand metric tons of CO_2, sell its excess right to pollute, and so make enough money to support the new technology.

In this case, because the excess right to pollute is used by someone else, you don't get as much greenhouse gas reduction as if the company was just forced to implement the new technology. But the virtue of a good cap-and-trade system is that it can incentivize extra reductions, as emissions caps are strengthened over time. Most cap-and-trade systems include a downward ratchet, where allowable emissions are automatically decreased over time by ever stronger regulations.

In 2005 The European Union launched a cap-and-trade system that ended up failing miserably. The price on carbon was too low, so no one wanted to buy the emissions permits, and the permission to emit was given way too freely. California watched this disaster unfold, and when it instituted its cap-and-trade system in 2006, as part of the AB32 landmark climate bill, it ended up with a system which has been working quite well. Under AB32 80 percent of reductions come from strong regulations, which is the most powerful way to get reductions. But a small amount of trading is allowed to happen, which has driven innovation in companies. And the money generated through this system is invested in equity initiatives which help reduce the impacts of greenhouse gas in low-income communities of color, such as public transportation initiatives. In 2018 that money came to $6.5 billion.[1] A well-run cap-and-trade system can be a part of a larger equity-based emissions reduction strategy.

Offsets

A policy tool that the climate justice movement hates even more than cap and trade and carbon taxes is offsets. Most people hear about offsets as a personal environmental choice, but in fact most of what is happening globally with offsets is at the policy level.

As a personal choice, when wealthy people fly, they often buy carbon offsets. If my flying from LA to New York produces one metric ton of CO_2, I can pay for someone to plant trees that suck up that same amount of greenhouse gas, thus offsetting, or neutralizing my impact. There are many companies that will sell offsets to individuals. There are a few reasons that climate justice activists are so critical of personal offsets. It might be that buying offsets clears my conscience, which might then allow me to fly more and not bother advocating for changes in the laws. And in many of these schemes the money does not really add additional trees. They often give money to projects that were already planting trees. For an offset to be legitimate, its benefits need to be in addition to what was already going to happen. People call this "additionality." Many people compare carbon offsets, especially individual ones, to the indulgences the Catholic Church used to offer to allow people to pay their way out of their sins.

Most of the global conversation around offsets is not about individuals buying offsets to lower their personal carbon footprint. Instead, most of the conversation has been about things that happen at a larger scale. Often when large companies produce greenhouse gases, in enterprises such as power plants, they are allowed to offset the emissions they create by planting trees, protecting forests, or paying for some other environmental good, somewhere else in the world. When money from polluters is given to offset their pollution, it is very difficult for the power relations involved to not lead to policies outcomes that don't really serve the communities they are supposed to serve. There is tremendous opposition to offsets from the climate justice movement in the Global South who have seen offsets used to promote projects that have been harmful at the local level or where the money simply went into the hands of corrupt officials.

Green Development Funds

In so much of the world, people live in poverty and have never had access to electricity, clean water, public sanitation, or quality transportation. It is not fair to simply say that these countries need to reduce their greenhouse gas emissions, without also finding ways that they can also use more energy. The UNFCC global climate talks have acknowledged that these countries need to "develop," meaning bring people out of poverty. The wealthier nations of the world will need to provide enough resources and technology to allow people in these countries to move out of poverty in ways that do not increase emissions.

The good news is that as green technology has improved, it is becoming increasingly clear that many countries in the Global South can skip the dirty phase of industrial development and move straight to sustainable development. Countries with deep poverty can leapfrog over industrial age technology, and electrify for the first time using solar power and electric systems of public transportation. Just as many countries never needed to install phone lines and went straight to mobile phones, so too they can move directly to micro- and smart grids for distributing renewable energy and not develop large coal-fired power plants and wasteful electricity grids.

The challenge for sustainable development in low-wealth countries is finding capital to invest in creating these green projects. The organization EcoEquity has been arguing for years that it is the responsibility of the wealthier countries of the world, that have contributed 79 percent of the total greenhouse gases in the atmosphere, to pay for these developments.[2] When we combine that fact with the reality that the worst impacts are being felt in the countries which have contributed the least to the problem, it is clear that the wealthier countries of the world have huge responsibilities to pay for sustainable development.

But how that money is transferred is important. Many countries in the Global South have governments that are corrupt, and "development" money can end up in the wrong hands. Many forces in the Global North try to profit from these transfers and so promote projects that are in their economic interest, not in the interest of the people who live in the countries the money goes to.

At the UNFCC meeting in 2005, nations from the Global South pushed to have a mechanism where the wealthier countries would pay for a development agency that would give countries money to pay to protect forests. The program developed out of that process was called REDD—Reducing Emissions from Deforestation and Forest Degradation. What seems like a good idea on the surface, paying countries to protect forests, has had huge negative impacts, especially on indigenous communities. The power relations between people who live in forests in the Global South, the governments of those Global South countries, those of wealthy countries, and the World Bank and other development agencies controlled by those wealthier countries, have generally meant that these schemes have largely been counterproductive.

REDD puts a price on the forest, and it often encourages countries to turn forests into nature preserves that don't allow people to live in them. It turns out that these depopulated forests are much more vulnerable to abuse from illegal loggers and miners than are forests which are inhabited and protected by indigenous communities. All around the world many indigenous communities live in forests, and if they are given power to control their forests, they are the people most likely to manage them in ways that are good for the climate.

A new version that tried to account for the problems of REDD was initiated in 2010 at the UNFCC meeting. The Green Climate Fund requires wealthier countries to pay into a fund that is then distributed to countries that need help with investments. The wealthier countries of the world have wanted to involve private capital in these projects. Many in the Global South have argued that these outside investors tend to fund the wrong projects. Instead, they are pushing for the funding to be controlled by local grassroots organizations that know which projects are likely to make the most impact. Unfortunately, these projects that are well integrated into local social fabrics tend to not be the ones that are exciting to foreign capital. Many climate justice advocates are arguing for an international Green New Deal, which invests significant resources in real solutions that work well to strengthen the local social fabric in ways that simultaneously solve environmental problems. The next iteration of green development funding will need to be even more mindful than the Green Development Fund of the power dynamics that make these projects very difficult.

Subsidies

As of this writing, governments are still giving money to support the fossil fuel industry. The US government paid for research that allowed BP to drill the deepest oil well in history, over ten thousand meters deep, in the Gulf of Mexico. The research paid for an extremely dangerous well, because the industry was interested in expanding its options for drilling in new kinds of places. That research led to one of the worst oil spills in history, as the well gushed for months. Just getting those subsidies to stop would speed the transition to a clean energy economy. The US government also continues to subsidize meat production with its support for growing cheap feed corn.

On the flip side, much of the work that has led to clean technology has been helped along by government subsidies. In the US, the government has given tax breaks to buyers of electric cars, to encourage adoption of them before it is really convenient or cost effective. This has helped tremendously with the transition to viability for electric cars. Many solar energy companies receive government investments to develop their technologies. The idea of subsidies is that they help stimulate for-profit companies to invest in the development of useful technologies and sustainable products. Much innovation can come from the engineering done by private companies. As long as the subsidies are well controlled and are not simply give a way to private corporations, they can be very helpful.

Elon Musk has benefited from $5 billion in subsidies. He is often seen as a business genius, but part of his genius is his ability to get public dollars for his projects. With public dollars should also come public control. Tesla has not been friendly to union organizing at their plants. And Tesla cars were designed to use a different charging plug than other electric cars. Tesla put massive investments into charging infrastructure. But that infrastructure can only be used by Tesla cars. If the government is going to subsidize private companies to develop for-profit products, it is important that there be strong strings attached to those investments, such as using standard plugs that benefit all electric cars and providing a union-friendly environment in factories.

Lawsuits

One powerful mechanism of environmental policy is the use of lawsuits. Legal action can help change the rules of the game and how those rules are applied. There are multiple lawsuits moving through the US courts against the major fossil fuel companies. Some allege that fossil fuel companies lied to investors about the financial viability of their companies in a future where policies are likely to make selling their assets less possible. Some suits argue that the companies are responsible for the damage their products cause. These suits are good for stigmatizing the fossil fuel companies, which can help lower their ability to control public policy. If the companies are required to pay billions in damages, those fees can also be used to invest in the transition to a sustainable society. And they might help move them to the brink of insolvency, which would help undermine their power.

Lawsuits are effective mechanisms for bringing problems to public attention, and they can help change laws and shift resources. There are many environmental lawyers who work on a contingency basis. They work for free in the hope that if they win their fees will be paid by the loser. This mechanism can make it possible for local communities to use lawsuits to shape policy.

Divestment

Divestment has been a significant tool of the climate justice movement. The goal of divestment is to change government policy. By divesting, we are stigmatizing the fossil fuel industry, and thereby weakening its ability to influence our political systems and to block needed climate legislation. The aims of the fossil fuel divestment movement—like antiapartheid, antitobacco, and other divestment movements before it—are to raise awareness, stigmatize a powerful political opponent, and win changes in government policy. In 2013 a study was produced by researchers at Oxford University, commonly referred to as the Oxford study, which showed that divestment movements have been successful in stigmatizing their target companies and ultimately in winning government action on their goal. Divestment was a crucial part in the struggle against apartheid in South Africa.[3]

Fossil fuel divestment was largely sparked in 2012 by an article published in *Rolling Stone* by Bill McKibben, "Global Warming's Terrifying Math." McKibben showed that the major fossil fuel companies have trillions of dollars' worth of carbon assets on their books. Their value as a company is based on the presumption that those assets will be burned. If regulations change, and those companies aren't allowed to burn their oil, gas, and coal, then those companies are overvalued. This means that these companies cannot be counted on to be part of the transition to a clean energy economy.

After publishing the article, McKibben went on a national tour called "Do the Math." The organization 350.org set up Fossil Free, a website with all of the resources necessary to get institutions to take their money from the top two hundred fossil fuel companies.

Fossil fuel divestment has been a powerful strategy in the fight against climate change. Before McKibben's article, many people working on climate change noted the ways that everything needed to change in society and how we are each of us as consumers is responsible for the problem. But if everyone is responsible, then no one is responsible. The fossil fuel divestment moment focuses our attention on the fact that 60 percent of all global emissions to date have come from just twenty companies.[4] And those companies are the single greatest obstacle to making the transition to a sustainable economy. The fossil fuel divestment movement is picking up steam as of this writing and has already achieved an important goal. It has helped the rest of society focus on fossil fuel companies as a major obstacle to making a just transition to a sustainable society.

Right now, there are hundreds of divestment campaigns in the US as well as globally. Over $12 trillion worth of asset portfolios have had fossil fuels divested from them. And as a result, the world is increasingly focusing on those companies as the culprits for slow action on making the transition to a postcarbon economy.

For the past two years I have been involved with a fight to get the California State Teacher's Retirement System (CalSTRS) one of the biggest pensions funds in the world, to divest from fossil fuels. In the past ten years, those investments have lost the fund over $5.5 billion, and those losses are projected to get worse. When we engage with the staff at CalSTRS, their strongest reason for not divesting is that they are doing important work by engaging with the fossil fuel companies.

There are activists all around the world who buy stocks in order to attend company shareholder meetings and advocate for changes in company policies. Experts in investment argue that you can get a company to change some things through shareholder engagement. You can often get better diversity policies on a company's governing board. But it is not a way to change that company's fundamental business model. For this reason, engaging with fossil fuel companies to make them sustainable is a losing strategy.

Investment

To address climate change, we clearly need massive development of solar, wind, and other clean forms of energy. And we need funding for public transit, and energy-efficient, affordable housing, particularly in low-income communities. We need to build the political will to have these investments supported by legislation and government funding. The concept of investment can be confusing because there are three main types of investment. Each of them is a different type of policy tool. They are: financial, private capital, and government investment.[5]

Financial investments are the purchasing of stocks, bonds, and other investment instruments, such as real estate.

These are the typical investments made by college and university endowments, and by pension funds. Financial investments result in changes in ownership of a share of a company, a bond, or land, but do not create new economic activity as do capital investments. Financial investments are made with the expectation of steady, fairly secure, but modest returns. They involve much less risk than capital investments because returns are tied to the profitability of an entire company and the general state of the economy, rather than to the success or failure of a specific business venture, such as a particular factory producing a particular type of solar panel.

With private capital investment, actual buildings, supplies, and other assets are purchased for a particular business venture such as producing solar panels, with the expectation of future profits. Private capital investment is highly risky in that it may produce huge returns or huge losses if a venture fails. Government subsidies can make it more attractive for private capital to invest in a particular technology, such as solar panel development.

A third type of investment is government funding. This is spending for such things as education, infrastructure, or energy research, in which there is no expectation of a financial return, and often no asset owned by the government as a result of the spending. Government funding can be for project that have a low rate of return, and they can be invested in things that don't turn a profit at all. Developing the projects needed to make the transition to a sustainable society requires investment, but those investments don't need to come from private funds and don't necessarily need to be profitable. They do need to be targeted at the things that need to happen that require capital to get off the ground.

Conclusion

There are many different tools for getting the world to net zero greenhouse gas emissions. And there is no reason to polarize and be critical of those who are working on one tool when you are working on another, unless that tool is actually counterproductive. The climate change movement, like most social movements, is a vast interconnected ecosystem that allows for all sorts of different species of action. When the ecosystem is healthy, those forms of action are synergistic.

CHAPTER 7

Finding Your People and Your Practice in the Ecosystem of the Climate Justice Movement

For a number of years I was involved in a fight to stop the expansion of a stretch of California's legendary Highway 1 that goes through the small town of Pacifica, where I live. The state transportation agency, Caltrans, wanted to double the width of the highway to facilitate development farther down the coast and to allow for more single occupancy vehicles. Our town does have rush-hour traffic, and the widening, after five years of slowdown from the construction, was supposed to make a small dent in that traffic. But it is common knowledge that when you increase highway capacity, you also "induce demand." If driving is easier, then more people drive. After a while traffic gets worse again. And many of us saw shutting down the expansion as part of a move away from single-occupancy vehicles, and thus as important for the climate.

While I trudged away in that many-years-long fight, wondering if I had set my sights high enough, I heard about the movement in Britain to not allow any increase in road capacity. In most of the Global North, more roads are rarely needed; what is needed are strategies for getting single occupancy vehicles off the roads and developing systems of public transportation that really work. It helped me immensely, in my work in my small town, to see it as part of a larger movement against highway building. By winning in Pacifica, we would send a message to the state of California that we needed to develop a new paradigm. The joy I felt when we won that struggle was magnified by my sense that our work was part of a bigger whole.

We should never underestimate the value of the small things we are doing because they aren't the magic bullet to solve the climate crisis. One of the worst things you can do to as a person involved in social change is to denigrate the importance of what you are doing. Even worse is denigrating what others are doing. But as in my example of the highway project, how I felt about my work was deeply connected to the narrative frame through which I understood it.

It is important that we keep in our minds the reality that we are a part of a team of hundreds of millions; that when we act smartly we are adding our grain of sand to the pile that is transforming society; that the movement is a grand ecosystem made up of a wide variety of elements, all working together in unpredictable and not always knowable ways. Deciding where to engage for yourself, and how much of yourself to dedicate to this work, is a very personal choice. This chapter is intended to help you understand the options you have for engagement and the considerations to make when looking for your place in the ecosystem that is the climate justice movement.

The Climate Justice Movement Ecosystem

In his book *Blessed Unrest* Paul Hawken describes the twin and deeply interlinked global movements for social justice and the environment, and sees them through the metaphor of the human immune system. These movements are growing as a response to the horrors of the exploitative economy which has increasingly ruled our world for the past five hundred years.

Explaining that metaphor, Hawken explains that the human immune system is actually a network of dozens of different strategies that the body uses to keep the integrity of the organism.

> The immune system is the most diverse system in the body, consisting of an array of proteins,

immunoglobulins, monocytes, macrophages, and more, a micro bestiary of cells working in sync with one another, without which we would perish in a matter of days, like a rotten piece of fruit, devoured by billions of viruses, bacilli, fungi, and parasites, to whom we are a juicy lunch wrapped in jeans and a T-shirt.... Just as the immune system recognizes self and non-self, the movement identifies what is humane and not humane. Just as the immune system is the line of internal defense that allows an organism to persist over time, sustainability is a strategy for humanity to continue to exist over time.[1]

Hawken argues that there are millions of organizations and tens if not hundreds of millions of people, doing that work. They are networked in complex and often loose ways. But they are all taking on the various forms of the disease that is destroying the livability of our home.

When we understand our work as part of an ecosystem, it helps us to see that the small elements are all related, that there is work to be done in a vast array of places, and that the work is most powerful when there are synergies between its different parts. We call that process a *social movement* when action toward solving a social problem begins to take off and proliferate in a wide variety of places and forms of action, and when the actions spread in ways that go beyond the control of any given organization or any given approach to solving the problem.

Social Movements

When social movements take off, they become a vortex that pulls in and stimulates a vast range of forms of activity. Social movements require a wide variety of kinds of actions: from people who get elected as politicians, to those who get arrested in civil disobedience actions, protesting those politicians. Effective actions take place along a wide spectrum of levels of commitment, from those who sign petitions to those who dedicate every waking hour to the movement. One important principle is that in a well-functioning social movement there is synergy between different forms of action.

Sometimes in movements people engaging in one form of action are so invested in their own virtue for doing what they do that they look down on the things done by people who have chosen a different form of action. To take one example, the pressure put on a system by outside actors engaging in civil disobedience can create openings that inside players, the people working closely with politicians, can turn into concrete policy victories.[2] Civil disobedience or lobbying work are tactics, or tools, in the organizer's toolbox. They are not strategies. A strategy is a clear plan to achieve one's goals. A good strategy usually requires a variety of tactics all working in synergy, even if they aren't always coordinated. Movements are made up of a vast range of people engaging in sometimes coordinated, but usually in autonomous, strategic activity.

Our movements are strongest when we value the wide variety of forms of action that are needed. Social movements are organic wholes made up, usually in messy ways, of a variety of organizations all working toward similar goals. Sometimes the organizations in a movement come together in a formal way and form a coalition, which is a grouping of organizations that work intentionally together to achieve a common goal. But often, movements are less intentional and more spontaneous than that.

Finding Your Place and Welcoming Others
Often when people join movements, they begin with something very small that seems like it isn't worth the time. But along the way they may form relationships, learn the terrain, and then slowly find a better place to put their energy. As you begin that process, there may be so many choices of where to put your energy that you end up feeling overwhelmed by the possibilities. That can lead you do to so much that your work

is not personally sustainable. Or it can lead to paralysis where you don't do much because no choice feels like it is the right one. A part of doing organizing work is finding a healthy and sustaining place for yourself within the vast ecosystem that is the movement. That process may be slower and more frustrating than you wish, and it may take more time than you hope. But the destructive patterns that are embedded in society have developed over centuries, and have built into the social fabric mechanisms for reproducing themselves. The world was not constructed to make transforming it easy.

Feeling guilty that you are not doing enough or doing the right thing is not helpful to the movement. That guilt is likely to make you want to stop doing anything. And for organizers, it is deadly to the movement to try to motivate people by pressuring them and making them feel that they need to do more than they feel comfortable with. It is crucial that you focus on developing a sense of accomplishment, for yourself and others, that comes from engaging in climate justice actions.

It is helpful to remember that you are part of a larger team and that your joy in the work will help bring others to the work. Feeling that joy is actually a productive part of the work. Martyrdom is very unattractive. People who try to be the most righteous activist are actually counterproductive to the work. People who try to pressure others into doing more than they are ready for are also counterproductive to the work.

If you are very busy and already overwhelmed with life but feel compelled to contribute to the transition to a sustainable society, it is possible to volunteer a few hours a week from home. Many organizations have small jobs that they can peel off for you. Hopefully what you get from that is a sense of purpose and you are truly helping. One thing you may miss out on, though, is the sense of connection that comes from being part of a movement. A well-run organization might be able to find ways to give a volunteer who does a few things a sense that they are appreciated and that their work is a significant part of the larger whole. But it is difficult to gain that sense of being a part of a movement without really working in community with others.

Any movement work is very personal and the qualities of the relationships really matter. I know people who are very conflict averse who find their home in organizations that are very well run and offer a lot of appreciation to their volunteers. I know other people with a thicker skin, who can forgive an organization that forgets to thank volunteers, that changes the plan a lot, so that a person sometimes does work that then ends up not being used. It is important that you really reflect on what your own personal needs are in terms of the kinds of people you like to work with, and what you need to feel good in your work. Finding your place in the movement involves that sort of personal reflection on the kinds of things you like to do and the kinds of cultures you want to be a part of.

It is also a good idea to think about the impact the work you are embarking on is likely to have. For those of us who love the thrill of action when it gets a bit chaotic and risky, nonviolent direct action can be a great tactic. But it is important to see it as a tactic rather than as a strategy. Some people chose to do it because it can feel like you are being strong and brave and setting a firm limit on what is wrong with society. But it is good to engage in tactics because we think they will make a difference, rather than just because we like doing them.

A tactic is the tool you use to achieve a larger goal. A strategy is the overall path of using a variety of tools to achieve a larger goal. Having a demonstration is not a strategy. The work done to get ExxonMobil to lose its political power is a goal, a particular divestment campaign is an issue to organize around, and organizing around that issue requires a strategy, which will involve a variety of tactics. Sometimes when people find a tool that really works for them, such as nonviolent direct action, they begin to want to use that tool for everything. It is always important to keep your strategic hat on and ask how what you are doing is supposed to achieve the goal you have set out, and to pick your tactics accordingly.

Inside versus Outside

One of the most basic divisions to think about when deciding where to put your energy is whether you prefer inside work, where you work within political systems to get them to change, or whether you prefer outside work, where you pressure those systems without directly engaging with them. Most work involves a combination of the two, but understanding the difference is helpful for understanding the layout of a movement.

There are many forms of outsider action. What they have in common is that they don't attempt to negotiate with power holders. Activists who use nonviolent direct action, who engage in actions that put themselves at risk of arrest, are working toward one end of the spectrum of outsider politics. Others forms of outsider work are protest action and guerrilla theater. Protests draw attention to a problem, and force the rest of society to take sides. Protests are also amazing ways to inspire a sense of community and power among people doing the work. Guerrilla theater is where one does something like a die-in, where they are not risking arrest but are doing something creative and surprising to get people's attention and make them think. These can all be fun and exciting and can help to shift the culture and its sense of urgency, and shift people's sense of what it will take to build sustainable society.

One of the great things about doing outsider work is that one is generally working with like-minded people, and can usually act in ways that are comfortable for you as you are. And by being part of a collective of like-minded people, one can feel an incredible sense of solidarity and clarity about the justice of what one is doing. Doing outside work, you don't need to compromise very much, you can be bold and radical. And you can wear what you want.

As of this writing, nonviolent direct action is increasing as a part of the climate justice movement. There are many reasons why this sort of action might not be right for everyone. And no one should act as though that is the only significant thing to do. Some of us are undocumented. Some of us can't afford to miss a day of work. Some of us have social anxiety. For some of us, being in the carceral system triggers past trauma. No one can choose what the best form of action is for another person. It is important to really think about who you are, to know what works for you.

When you work from the inside as a politician, in an electoral campaign or doing lobbying work, you get to be where the rules are made. And this work is crucial. Many of my students helped to elect Ash Kalra, a California assembly member who has gone on to pass several bills that have been good for the climate. He is an amazing climate warrior, as are those, including my students, who helped get him elected. Entering those spaces often requires personal compromise in terms of dress and how we act in order to be effective. When we all dress up in professional attire and talk to our elected officials or knock on doors as part of an electoral campaign, we can make a very significant difference. Some people enjoy doing that, and it is crucial that some people do it.

This is one of the places where the more radical people in a movement sometimes make a mistake. They find it important to stay grounded in their alternative view of how the world should be, and so see it as selling out or as an unethical compromise to act like the people we might be opposing. But in the complex ecosystem that is our movement, there are many things needed to make the changes we need. One can choose to never be that person playing the inside power game, but it is not helpful to the movement to denigrate the people who do that hard, inside work.

On the other side, people who play an inside game can sometimes demean the work of outsiders. Insiders spend time with powerholders. Those inside players don't like pressure and will often say that protestors are ridiculous and don't understand how change happens. And the inside players on our side can sometimes find the criticism of the work of outsiders that comes from the politicians being pressured puts stress on their relationship with those powerholders. Sometimes insiders have carefully crafted a message and outsiders who come to protest can disrupt that work. But the reality is that the radical outside politics usually helps to shift the balance of power by changing the culture in ways that create openings for the inside players.

The insiders often think that the outsiders are making them look bad a not serious. The outsiders often see the insiders as sold out and too willing to compromise. It is important, and not always easy, for people on either side of this divide to understand their different roles. Our movements are stronger when there is coordination between those doing insider work and those doing outsider work. In a well-functioning movement, even when there is not direct coordination, there is synergy between these approaches and there are a lot of people engaged in each.[3]

Respecting Movement Subcultures

Movements are made up of a wide variety of different segments, and each segment may have with in it a different subculture. It is important that when we look at moment, we respect the very different roles that different parts of a movement play in our interconnected ecosystem. Sometimes people of color or queer people will want to organize with their own groups, so they can contribute without having to fight against the oppression dynamics they might experience in a more multifaced group. Or they may form a separate group simply because they might find more joy in such a group.

One subsector of the movement that is growing at a rapid rate as of this writing is the climate youth movement. Joining the youth climate movement, rather than joining adult organizations, can be a great choice for a young person. Young people's voices are carrying a lot of weight in the present moment and keeping those voices elevated is important. There is something powerful about the voices of young people saying to adults in positions of power that they have not been using their power responsibly and that those adults in positions of power need to urgently do all they can to stop the burning of fossil fuels.

Another virtue of the separate climate youth movement has been its consistency in calling for a just transition and in its attention to the social justice aspects of the climate crisis and of the movement itself. Youth have been better than the adult movement as a whole at insisting on a climate justice lens through which to view the problem as well as the solutions. If the youth were working in adult organizations, that powerful and distinct voice might be watered down.

Youth-led organizations are popping up at an incredible rate, and the youth have done an amazing job networking and engaging in a wide variety of actions in coordination with adults, while also keeping their distinct voices and approaches to the work clear. It is crucial that adults who work with youth remain mindful of the power dynamics in those relationships and continue to amplify the voices of youth rather than trying to co-opt or control them. And it is important that youth in the movement keep clear about what is different and crucial in their work.

Sectors of Climate Work

While many people think of climate justice work as taking place in the streets, that is partly because that is where our work is the most visible. But a healthy movement ecosystem has many different sectors where important work takes place.

Service work around climate resilience and displacement

For some people, the sense that they are directly impacting others and helping them is important to their sense that what they are doing matters. After climate disasters, many local people engage in direct aid to rebuild and to provide medical and other help to survivors. Often that work helps expose gaps in the social safety nets and can be a basis for demands for social transformation.

Mental health work

There is a lot of work to be done in small, self-organized groups of people gathering to support each other in their mental health as we experience the terror of what is coming. One can organize such a circle, participate in one, work as a therapist, or promote robust mental health practices within other places and organizations.

Artwork

Every movement needs its artists. Social movements are inspired and given clarity, and they communicate with the broader world through creative expression. Sometimes artists form collectives to support each other in their connections to the movement. Others work embedded in broader organizations. Still others work on their own and their works circulate in ways that are inspiring, educational, clarifying, healing, and attention-grabbing.

Educational work

Some of us are teachers and find our place in the movement by giving workshops, by offering formal and informal classes, and by working to transform education systems.

Policy advocacy and electoral work

Important social changes happen when social policies are changed, and that work requires people who do research on alternatives, lobby politicians, work to elect climate warriors to office, and develop and propagate policy proposals.

Journalism and media activism

All movements need journalists to amplify our messages, to spread our ways of understanding the world, to expose the actions our opponents, and to help us understand what is going on. Media activists can be staff members at news organizations. They can be freelancers. They can be bloggers. They can be social media influencers. All of these locations are places where important media work can happen.

Jobs in the Climate Justice Movement

For many people, the best way to make an impact is to get a job in the climate change movement. It can feel very satisfying to spend one's whole working life doing things that make a difference. For many people this means working in engineering or city planning, or for an advocacy group. For many jobs the best major is environmental studies or sociology, communications, political science, and a wide variety of other majors. In some places it is possible to study community organizing.

Many of those jobs pay well, and many of them allow you to do a lot of good work. It can feel great to go to work every day and help change how cities manage their garbage, work on lobbying for impactful legislation, designing the next generation of tools for sustainability, or educating young people about gardening.

The hard part of doing social change work through your job is that you may not be able to do the kind of work you find most satisfying or driven to do, or that you see as the most important. It is very unlikely that you'll find a job that asks you to challenge power or engage in disruption. For myself, I decided to be a community college teacher, so that I could do something socially valuable as my job and still have enough spirit left over at the end of the day to engage in my more disruptive forms of activism. Over time, my job has changed and now I teach and mentor young people to be agents of social change. But I was at my job for almost twenty years before the two halves of my life came together.

Many people choose a career that that gives them the stability they want and that leaves them with enough left over to be activists, and their activism is not at all related to their jobs. That path allows them to freely choose what kinds of climate action to engage in.

Doing Your Own Thing or Starting Your Own Group

There is a lot you can do on your own, especially if you are careful to make sure what you do helps support larger strategies. If you decide that a company is bad and you decide to not buy their products, it is not likely that your work will add up to significant, meaningful change. The logic behind that action is the market logic that says that markets respond to our buying or not buying. As one individual consumer, if you aren't joined by many others, your action isn't likely to make much of a difference. Joining organized boycotts is a whole different thing. Thoughtful work to organize a boycott can be a big step to making a significant difference. Organizers can be strategic about picking a company that is likely to be vulnerable to pressure and to be able to work strategically to get a win.

That advice holds for many forms of action you can take on your own. If organizations are working on a bill, you can do your own individual work to pressure your local representative to support or oppose it. If there is a boycott or divestment campaign, you can add your voice through letters or talking with people you know. The more your induvial work helps add energy to some work thoughtfully and strategically organized by others, the more likely it is that it will be effective. And a similar logic follows for starting your own group. You can get together with a group of friends to do a variety of actions, like writing letters, raising consciousness, and in a variety of ways adding to larger goals being worked on by other organizations.

Almost every town or city is a good target for climate change action. If your town doesn't have anything going on, you can inquire about whether the town has a climate action plan and push it to have one, or to make an existing one better. The same holds true for most local institutions.

Does your school have solar panels, environmental education, bioswales, good equitable transportation? Is your place of worship invested in fossil fuels? There are good targets everywhere.

Conclusion

It may take you a while to find the work that is right for you. It is worth experimenting and asking a lot of questions until you find the work that is impactful and that helps you to feel a sense of meaning and purpose and joy in your life while also making a difference in the climate crisis. The climate crisis is real, urgent, will take all of us doing our best to avoid the worst outcomes.

The climate crisis hit at a time when many governments were in a phase of deregulation and allowing markets to make important social decisions. It hit at a time when many in the environmental movement liked to focus on green consumerism, and some propagated the image of wealthy white people as the model for environmental action. The climate crisis hits all of the wrong buttons for the human psyche: it seems far off in place and time, it seems like everyone and no one is responsible for it, and looking at it triggers a sense of guilt for many people. For all of those reasons we have been dealt a very difficult hand, yet we have no choice but to play it. And playing it requires that we use all of our intelligence and courage.

And we will thrive best under these hard circumstances when we think hard about the best places to put our energy. We need to be selfish in making sure that the flame of our commitment is increased rather than decreased by our work. We need to each find a path of engagement what will allow us to stay in the fight for as long as it takes for us to achieve our goals. That will require resilience and persistence, and it will require us to find joy and sustenance for our lives in our climate action.

CHAPTER 8

Self-Care

I have to admit that as involved as I am with the climate justice movement, when people start to talk to me about the new news they heard about why it is all worse that we think, I tend to close my ears. I know that it is important that people know all of that. And it is crucial that we help the general population to find pathways to effective action as they become increasingly aware of the urgency of the crisis we are in. But I also know that panic doesn't actually drive action. And I know that for myself, I am motivated enough to do a lot of work, and to find a place in my life for that work, and so I would personally rather not spend time facing those hard truths.

One of my favorite expressions is "pessimism of the intellect and optimism of the will." It comes from the Italian philosopher Antonio Gramsci, who argued that it is important to take a sober and realistic look around and understand the situation you are in, to not sugarcoat it or minimize its dangers. And at the same time, he argued, it was important to always evaluate any situation for where the opportunities are to make a difference and to focus one's attention on those possibilities.

That quote is my most fundamental mantra in life. I focus on the cracks I see before me where significant action is possible. And I try not to think much about all of the people whose approach to the crisis is frustrating and annoying, those things that should be paths forward but are not, the people who aren't doing anything, and just how pathetic the mainstream media and mainstream politicians are. One of the lessons people learn in mindfulness workshops is that you can train your attention to be on the things that matter.

Those lessons are in line with the overly famous Serenity Prayer from radical theologian Reinhold Niebuhr: "God, grant me the serenity to accept the things I cannot change, courage to change the things I can, and wisdom to know the difference." Whether or not you believe in God, there is an important piece of insight in that prayer: that we can chose to focus our attention on the places where we can make a difference. I am happy to listen to real information about the new bad news about the crisis. But I have no patience for people who just want to spin their wheels in that sandy rut.

It is a delicate balance for all of us engaged in climate action to find a place where our work supports rather than destroys our mental health. In 2017 the American Psychological Association put out a report called *Mental Health and Our Changing Climate: Impacts, Implications, and Guidance.* The authors note increases in anxiety and depression for everyone facing the climate crisis. And for people who have been through extreme weather events, such as a fire or hurricane, there is often serious post-traumatic stress disorder. We all have different ways of dealing with hard things. And whether one is involved with climate action work or not, whether we are in denial or open to the fullness of the catastrophe we face, we are all in some ways dealing with the hard reality of the climate crisis.

Action Can Help

I have found that action helps my mental health for a few reasons. One is that I am especially allergic to denial. I grew up in a home where hard things weren't mentioned. Part of my growing into a healthy adult has been a ruthless desire to name hard things. When I am in polite society where the hard aspects of our world are not talked about, whether it is mass incarceration, empire, or the climate crisis, I feel uncomfortable. I find it deeply disturbing to talk to people about their vacation plans and home remodels as if nothing were happening. Feeling a connection to the movement is infinitely better than feeling that the world is on fire and no one is noticing.

And the relationships one forms with others who are equally committed, is the biggest form of therapy I have going in my life right now. Being involved in the climate justice movement, one can find a sense of community with other people who care as much as you do, and who are living with the reality of the crisis embedded in their everyday understanding of the world. It feels good to me to be around people who actively share the realities of the traumas of our world and share also a passionate drive to do something about it.

Surround Yourself with Solutions

Being involved with the movement I am more likely to be aware of the huge things that are happening under the surface to build a just transition to a sustainable society. People not in the movement often think that nothing is being done to address the climate crisis. As we have seen in pervious chapters, that is far from the truth. We are in a time of accelerating disasters and of accelerating action to fend off those disasters.

The more people are involved with the movement, the more they are aware of all the things being done to reweave our social fabric. Those of us really in the climate movement tend to know more clearly than others just how catastrophic the situation we are in is. But we also know that there is tremendous movement happening at every level of almost every society in the world. That feeling of being part of such a massive transition is exciting.

Engage in Personally Sustainable Practices

But being involved with movement work can only function as therapy as long as the organizations we associate ourselves with have healthy supportive cultures. We need to think, as individuals and as members of groups, about what practices can keep us healthy in our work. It is crucial that we avoid martyrdom.

When I was involved with Central America work in the 1980s, someone very prominent in the movement told me that I shouldn't go to graduate school, and that if I really cared about the movement, I would dedicate my life to it. Years later I found out that he was living off a trust fund. For myself, it was important to get a career that paid a decent wage and would leave me time and spirit for activism. I work with a lot of motivated, passionate young people in my job, and one the things I tell the ones who are so passionate about their activism that they aren't making progress in school is that I want them in the movement when they are my age. That means putting their activism into a sustainable place in their lives. Climate action should not come at the expense of our family relationships, our romantic lives, the pursuit of economic stability, having fun, or especially our mental health.

Curate Your Media Feed

One part of maintaining our mental health is being mindful about how we consume information. Many people have news consuming habits that contribute to their unhappiness, without contributing to their ability to make a difference. If we aren't intentional about how we consume news, it is easy to be saturated with too much information, with misinformation, and with information that is overly sensationalized because that's what sells ads.

If all of your news comes from the "news" shown at the top of Google, you are allowing a corporation with no credibility as a news source to control your feed. Most social media platforms are fundamentally ad agencies, so most of them boost the visibility of things that excite strongly positive or negative responses. Consuming news that way can be emotionally exhausting, and it can inform us about things that don't matter very much. If most of your news comes from what friends share on social media, that is also likely, for different reasons, to skew toward the sensationalistic.

Just as an inspiring art show was curated by someone with a perceptive eye and a deep sense of what is interesting and important, so should we work actively to curate our news feed. Most of us know a few people who are more interested in particular subjects than we are and can be counted on to be insightful curators. Following those people in social media can help us develop a feed that gives us helpful and reliable information and does not compromise our mental health.

One can also get a good news feed by actively going to a well-respected news source. For almost any political persuasion there are sources that are sensationalistic, unreliable, and full of stories put there to flatter the baser impulses of their audiences. And similarly, almost every point of view has sources that rely on standards of good journalism, such as using credible sources, checking facts, and deep investigations. Finding reliable sources of news is important for being a well-informed global citizen. It is a good idea to sometimes read sources whose point of view challenges you to understand how other people see the world. It can be valuable to read both reliable and unreliable sources from a different point of view, to get a window into the minds of others.

But when we consume news to inform ourselves about what is happening in the world, there is nothing wrong with relying on sources that accord with our own point of view. None of us can be deeply critical and analytical every time we read or watch news. It can serve us well to find news sources that have actually invested time and resources into getting more information and understanding of those things that we would like to know about. And it is helpful to rely on those sources to expand our worlds into areas that are new to us.

One of the best sources of news on the climate crisis is InsideClimate News. Joe Romm blogs on the climate crisis and curates the climate page on *Front Page Live*. Authors who are great on climate activism are Bill McKibben, George Monbiot,

Winona LaDuke, and Naomi Klein. Some of the best podcasts for general progressive news are *Democracy Now!*, *Reveal*, and the *Intercept*. The *Guardian* newspaper, which has a free online app, has taken a very proactive approach to reporting on the climate crisis. Other good online news and analysis sources are the *Nation*, *Truthout*, and *Common Dreams*.

Knowing about the climate crisis can be demoralizing and exhausting. Knowing about the climate justice movement can be exhilarating an inspiring. Finding the right amounts and kinds of news to consume to keep you motivated to engage in climate action requires conscious choices about where to put our attention. There is nothing wrong with not always reading much news and with sometimes taking a break from information.

Critical Consciousness

As people become aware about what is wrong with the world, they often begin to see the hidden patterns and structures of power underlying social reality. For some people that first exposure to just how bad it really is can be devastating. To keep from staying devastated, it is crucial to learn about solutions as one learns about the depths of the problems. Challenging the mainstream, having the audacity to believe that one can make a difference, and committing time to challenging power, are not easy. People stay in that game and are successful in making a difference when they feel they are part of a movement ecosystem that is building a world that works for us all.

That's what happened for me. When I got politicized in 1980 around opposing US support for the dictatorship in El Salvador, I became engaged right away and, through that engagement, discussed politics, read a lot, and learned to analyze the world around me. I ended up simultaneously doing work to change the world, deepening my consciousness, and coming into contact with amazing people who became close friends and allies and who enriched my life.

The Brazilian philosopher Paulo Freire named the process of coming to see the problems of the world and simultaneously coming to see the possibilities for engaging in social change *critical consciousness*. Before coming to critical consciousness, people are likely to have an orientation to the world that takes things as they are as given, and positions themselves as following the mandates of society to carry out the roles that they are offered. This orientation to the world perpetuates systems of oppression. After coming to critical consciousness, in Freire's view, we see the structures of power, we see our own places in them, and we come to be agents capable of transforming society in ways that help us break the repetition of those entrenched structures of power.

Critical Consciousness under Capitalist Individualism

But what happens when we come to consciousness of those structures of power in a social system that create us as fundamentally disconnected from one another? Freire developed his work among nonliterate peasants in rural Brazil. The peasants whom Freire worked with were oppressed by capitalist economic practices, and governments that supported them. That system denied them access to the land they needed to survive. It policed the control of land through violence. But those peasants were not fully articulated into capitalist cultures of consumer consciousness. They lived in communities that were structured with deep patterns of domination but also with complex systems of community interdependence. They were not raised in a society that told them from day one that their purpose for existence was to pursue individual gain and happiness through buying things.

In a more thoroughly capitalist world, where our connections with one another are structured by deep patterns of disconnection, where the systems of meaning through which we perceive our places in the world have been completely colonized by consumerism and individualism, the problem of critical consciousness has a new set of challenges.

People becoming aware of social justice in twenty-first-century industrialized societies often experience themselves as profoundly disconnected from others. Coming to consciousness in a toxic culture of individualism and consumerism can make it difficult to find ways to engage in effective action. Effective action involves coordination with others to transform social patterns. It involves strategy. It involves action toward often far-off and elusive goals. It requires persistence when it seems like things are going nowhere and the work isn't especially gratifying. It involves relationships. All of those things require an investment of ourselves in collective work with others. And it involves seeing ourselves as members of something larger than our individualized self. Most effective work toward social change requires us to think strategically and act boldly. It is about what we do to transform social structures more than it is about the purity of who we are as individuals.

Systemic domination and the climate crisis both lead to trauma. When social change work is done right, being involved can be incredibly healing. Much social justice work involves putting our pain into a larger context of systems of domination and a world gone wrong, and working to change the things which have harmed us. Seeing oneself not as a victim, but as an agent of social transformation, can help us develop new empowering narratives of ourselves as part of a large-scale process of healing. And in that work, it is possible to heal oneself as one heals the world.

Social Media Activism and Toxic Self-Righteousness

Unfortunately for the development of effective action for a social justice, there is a tempting bypass to the hard but rewarding work of developing real relationships around effective action to heal the world. That bypass resonates beautifully with so much that is wrong with the world as we find it: the politics of competitive awareness that thrives in the world of social media activism.

I worry now that many people who are coming to a clear understanding of what is wrong with the world are not also getting swept up into meaningful action or into relationships that sustain them. Recently, I've met many young people who are very aware of the systems of domination in which they live but find the world of social change so full of meanness that they don't become activists. Or if they do, they leave after a few short years. Being aware and seeing what is wrong serves no good purpose if it is not also aligned to engagement aimed at transforming social patterns. And no one is going to stay in the work of social change in ways that make a real difference over the long term, if they don't find it nurturing and healing.

While social media can provide great tools for social justice work, in many social justice circles, and especially online, the world of social justice is in danger of it becoming a circular firing squad, where people fight to see who is the most aware, and where they see activism as primarily about challenging the lack of awareness in others. If we want to make real progress in fighting the forms of domination that are destroying our lives and the habitability of the planet, we need to find ways to support each other in learning how to work together for social change. Some of that has to do with having empathy for people who are just coming to consciousness. Some of it has to do with seeing what we are facing as related to institutional structures that need challenging, as opposed to simply being about identity and interpersonal interactions.

Many people who have consciousness of what is wrong with the world then use that consciousness as a cudgel to bash others with who aren't quite as aware as they are. They engage in what scholar and activist Elizabeth Martínez called the Oppression Olympics. By focusing on the ways that one is oppressed and calling out people who don't see the problems and know the language of domination as well as themselves, they gain a sense of righteousness that makes them feel good in the very short term. But that sort of righteousness does not contribute to healing their individual trauma or the social traumas that cause them pain.

And just as most people are targets of some form of oppression, we are also in a variety of ways carriers of privilege. I have found that as people begin to develop critical consciousness, doing a little bit of work to focus on the ways that one has privilege and developing a sense of humility based on that can go a long way toward cutting through the toxic culture generated by only focusing on how aware one is based on the ways that one has been oppressed.

Doing Politics *in Real Life*

Another important cure for destructive social justice awareness, is to take one's politics offline. In the #BlackLivesMatter and #MeToo movements we can see that there are forms of connection we have online which can be effective at spreading useful ideas, sharing strategies, and transforming cultural meanings. But so much of what is transformational in both of those movements has happened when the energies unleashed by those connections, which flow so quickly online, have been also captured and mobilized by people working in communities of practice on the ground to achieve common goals. The relationships which are formed in face to face work, which also involve sharing meals, sharing fun, and relying on each other for mutual support, are crucial counters to the feelings of alienation and experiences of cruelty that are routine online.

Building Empathy and Healing Relationships in Our Work

Many of us who have been targets of systems of oppression carry trauma from wounds of misrecognition, silencing, violence, and poverty. Social justice work is about creating a world where those things don't happen to people, where we transform the social systems that cause most of the trauma that people experience. Yet as we work to challenge the systems that create that trauma, we need to ask if what we are doing is simply playing out our pain. Or are we transforming the systems that cause that pain, and building a better world for ourselves in the present, and for others in the future. Doing that requires an investment of time and energy in building positive caring relationships with others who are also engaged in the effective thoughtful work challenging systems of domination.

Discussing her disillusionment with social justice communities author Kai Cheng writes:

> I became quite valorized in my local community as a "good" upholder of social justice because I was very good at using the right language and doing the right things, and I tended to apologize unreservedly and perfectly when I "fucked up." I now recognize this as a skill born of trauma: the ability to ceaselessly and accurately scan the people in one's environment for a sense of what will please them and to enact it, no matter the cost to one's long term health. this skill is a brilliant short-term survival strategy, as most trauma strategies are, designed to negotiate the unpredictability and cruelty of punishment—and unfortunately, much of social justice is deeply embedded in a punishment narrative.[1]

Cheng goes on to argue that it is crucial to base our social justice work on compassion and building a healthy world, as opposed to punishing people who are oppressive. She quotes Adrienne Maree Brown, who, in her book *Emergent Strategy*, writes.

> You have the right to tell your story. . . . You do not have the right to traumatize abusive people, to attack them publicly, or to sabotage anyone else's health. The behaviors of abuse are also survival based, learned behaviors rooted in some pain. If you can look through the lens of compassion, you will find hurt and trauma there. If you are the abused party, healing that hurt is

not your responsibility, and exacerbating that pain is not your justified right.[2]

In our movements we can work to develop healing cultures, which allow people to heal pain, rather than being places that reproduce our traumas.

Prefigurative Politics

The alternative to playing out our trauma by living online and trying to win the Oppression Olympics, is to focus on the bigger picture of the beloved world we are trying to create, and to try as much as possible to enact that world in our work to get there. Prefigurative politics is a concept developed in the 1970s by feminists who wanted to develop a politics that was not just about working miserably for a future utopia.[3] Rather, it focused on trying as much as possible to live those utopian values of care, antioppression, and sustainability, in our work to build a better world.

In our organizing, if we aren't careful, we can be reduced to worker bees who slavishly do what we do in order to achieve our goal of a world without too much carbon in the atmosphere. Doing work in that way can achieve many goals. But our work is stronger and people's commitments to it can be deeper, when we do the work in ways that don't replicate the miserable alienated labor so many people experience under capitalism. If we attend to the quality of our human relationships in our work, we can be happier and we can build the beloved world we are working toward with the steps we take in the present. If we keep in our minds the strong vision of the just, emotionally rich world we are working toward, then we can be more likely that the steps we take are actually steps toward that ideal.

Declare Victories along the Way

For people aware of just how much needs to be done quickly, anything we do can feel like it isn't enough. But if we only focus on what we haven't done, it is hard to do much of anything. We need to praise ourselves and those around us for the work we put in even when it doesn't bear fruit, and we need to claim victories when we make even small gains. Celebrating our wins with others builds our spirit and strengthens our commitments. Morale, appreciation, and hope are crucially important for keeping ourselves motivated. It is a good practice to end meeting with a round of appreciations, and to take time out regularly to express gratitude for ourselves and for others.

Conclusion

Finding one's place in the ecosystem of the climate justice movement can be deeply personally healing. Doing what we can as effectively as possible, and bringing others along into the work, are necessary for our survival as a species. The work we do in the next years will determine the future habitability of the planet. The period to come is likely to see more devastation as we work to transform our future. And it is also likely to be a period of incredible heroism and inspiration. As we have seen with the Covid-19 virus, social processes can move quickly from a state of denial and minimization, to a state of intense mobilization. Just as with the virus, the hope is that as the action accelerates, the forces working toward a better future prevail over the forces of destruction.

All around the world socially just and climate friendly transformations of all of the practices that undergird human society are being implemented. We are in the midst of a massive transformation in every country in the world. Engineers and scientists have done amazing work developing solutions. The United Nations has done a lot, under heavy counterpressure from the forces of business as usual, to get the governments of the world to commit to realistic goals, and to commit to a process to increase those goals. The fights to undermine the power of the fossil fuel industry and to stop it from being financed are accelerating.

Solutions to the crisis are being implemented all around the world and in every sector of society that needs action. Each of us needs to help increase the speed and intensity of work being done to implement those solutions and build a better world for all of us. There is a rich ecosystem of activism that is easy to join, and that ecosystem includes a huge variety of possible forms and levels of engagement. Finding one's place in that ecosystem and being part of the solution can be much healthier than sitting on the sidelines and fretting about how bad it all is.

This is the fight of our lifetimes. It is a global fight for human survival like no other generation in history has ever faced. And being a part of the solution can be a powerful and life-transforming experience.

Notes

Introduction

1   This quote comes from a speech Arundhati Roy gave at the World Social Forum in Porto Alegre, Brazil, January 27, 2003. The phrase "another world is possible" comes out of the Zapatista movement, which was very influential on the global justice movement in the 1990s.

2   Tim Dickinson, "Study: U.S. Fossil Fuel Subsidies Exceed Pentagon Spending," *Rolling Stone*, May 8, 2019, https://www.rollingstone.com/politics/politics-news/fossil-fuel-subsidies-pentagon-spending-imf-report-833035/. The *Rolling Stone* article relies on a study done by the International Monetary Fund: David Coady, Ian Parry, Louis Sears, and Baoping Shang, "How Large Are Global Energy Subsidies?," May 2015, https://www.imf.org/external/pubs/ft/wp/2015/wp15105.pdf.

3   Jane Mayer, *Dark Money: The Hidden History of the Billionaires Behind the Rise of the Radical Right* (New York: Doubleday, 2016), chap. 8.

4   The concept of "just transition" is attributed to Anthony Mazzocchi, president of the Oil, Chemical, and Atomic Workers Union, who used it to describe a transition to environmental sustainability that would fully compensate workers in declining industries and ensure that the new economy was one that worked for labor. See *Labor Network for Sustainability*, "'Just Transition'—Just What Is It?" https://www.labor4sustainability.org/uncategorized/just-transition-just-what-is-it/, accessed May 2, 2019. Also see Climate Justice Alliance, "Just Transition," https://climatejusticealliance.org/just-transition/.

5   Ira Katz Nelson, *When Affirmative Action Was White: An Untold History of Racial Inequality in Twentieth-Century America* (New York: W.W. Norton, 2006).

Chapter 1 What We Are Up Against: Science and Politics

1   Joseph Romm, *Climate Change: What Everyone Needs to Know*, 2nd ed (Oxford: Oxford University Press, 2018), 9–12.

2   Scientists measure other greenhouse gases in terms of their Global Warming Potential (GWP) as compared with CO_2. This is an inexact

science, since the different molecules last different lengths of time and have different impacts on the atmosphere. Because it is the chemical used as a reference, CO_2 has a GWP of 1. Methane has a GWP of 28–36. And many of the chemicals used for refrigeration have GWP numbers in the thousands.

3   International Institute for Sustainable Development, "Reforming Subsidies Could Pay for a Clean Energy Revolution," updated August 8, 2019, https://www.iisd.org/gsi/news-events/reforming-subsidies-could-help-pay-clean-energy-revolution-report.

4   National Network for Immigrant and Refugee Rights, "Fact Sheet: Climate Change, Global Migration, and Human Rights," September 2018, https://www.nnirr.org/~nnirrorg/drupal/sites/default/files/climate_change_and_migration_fact_sheet_final.pdf.

5   Romm, *Climate Change*, 159. Parts per million means the number of units per a mass of something as a percentage of a million units of that mass.

6   Mary Robinson, *Climate Justice: Hope, Resilience, and the Fight for a Sustainable Future* (London: Bloomsbury, 2019), 6–7.

7   Center for Global Development, "Developed Countries Are Responsible for 79 Percent of Historical Carbon Emissions," August 8, 2015, https://www.cgdev.org/media/who-caused-climate-change-historically.

8   Matthew Taylor and Jonathan Watts, "Revealed: The 20 Firms behind a Third of all Carbon Emissions," *Guardian*, October 9, 2019, https://www.theguardian.com/environment/2019/oct/09/revealed-20-firms-third-carbon-emissions.

9   Jane Mayer, *Dark Money: The Hidden History of the Billionaires behind the Rise of the Radical Right* (New York: Doubleday, 2016), chap. 8.

10 Some people don't like the concept of net zero, because a polluter can say that they are doing something positive to "offset" their pollution and say that means that their emissions are "net-zero." But if you don't use a concept like net-zero, then you only focus on emissions and not on carbon sinks. At least as important as stopping putting greenhouse gases into the atmosphere is working to protect and expand the processes—such as forest protection, tree planting, and regenerative agriculture—that sequester carbon, or pull it out of the atmosphere and store it in plants and soil.

11 Robinson, *Climate Justice*, 134–35.

12 Robinson, *Climate Justice*, 138.

13 Nicholas Stern, *The Economics of Climate Change: The Stern Review* (Cambridge: Cambridge University Press, 2007).

14 Cristina Figueres, in Johan Falk, Owen Gaffney, et al., *Exponential Roadmap: Scaling 36 Solutions to Halve Emissions by 2030*. Version 1.5, 2019, https://exponentialroadmap.org/wpcontent/uploads/2019/09/ExponentialRoadmap_1.5_20190919_Single-Pages.pdf, 31.

15 Tom Di Christopher, "Climate Disasters Cost the World $650 Billion over 3 Years—Americans Are Bearing the Brunt: Morgan Stanley,"

CNBC, February 14, 2019, https://www.cnbc.com/2019/02/14/climate-disasters-cost-650-billion-over-3-years-morgan-stanley.html.

16 CoastalChallenges, "44 Percent of Us Live in Coastal Areas," January 31, 2010, https://coastalchallenges.wordpress.com/2010/01/31/un-atlas-60-of-us-live-in-the-coastal-areas/.

17 Oil Change International, "Fossil Fuel Subsidies Overview," http://priceofoil.org/fossil-fuel-subsidies/.

18 ARO Watch. "Tracking the Trillion Dollar Retirement of Oil," https://www.arowatch.org/. Accessed December 24, 2019.

19 Union of Concerned Scientists, "2.5 Million Homes, Businesses Totaling $1 Trillion Threatened by High Tide Flooding," June 18, 2018, https://www.ucsusa.org/about/news/25-million-homes-threatened-high-tide-flooding.

20 Erin Durkin. "North Carolina Didn't Like Science on Sea Levels . . . So Passed a Law against It," *Guardian*, September 12, 2018, https://www.theguardian.com/us-news/2018/sep/12/north-carolina-didnt-like-science-on-sea-levels-so-passed-a-law-against-it.

21 Bruno Latour, *Down to Earth: Politics in the New Climatic Regime* (Cambridge: Wiley, 2018).

22 David Harvey, *A Brief History of Neoliberalism* (Oxford: Oxford University Press, 2007).

23 Latour, *Down to Earth*, 2, 9–10.

24 Latour, 11.

25 Latour, 23.

26 Latour, 66.

27 Latour, 105.

Chapter 2 Another World Is Possible

1 Paul Hawken, *Blessed Unrest: How the Largest Social Movement in History Is Restoring Grace, Justice, and Beauty to the World* (New York: Penguin, 2008), 2.

2 George Monbiot, *Heat: How to Stop the Planet Burning* (New York: Penguin, 2007).

3 Maisa Rojas, "The Climate Crisis and Inequality a Recipe for Disaster: Just Look at Chile," *Guardian Weekly*, December 13, 2019, https://www.theguardian.com/commentisfree/2019/dec/08/un-conference-global-heating-cop25-chile-madrid-climate-crisis.Guardian *Weekly*, December 13, 2019, 46.

4 Naomi Klein, *This Changes Everything: Capitalism vs. the Climate* (New York: Simon and Schuster, 2015), 7.

5 Robin Blackburn, *The Making of New World Slavery: From the Baroque to the Modern, 1492–1800* (New York: Verso, 1997).

6 Cynthia Kaufman, *Getting Past Capitalism: History, Vision, Hope* (Lanham, MD: Lexington Books, 2012).

7 Gar Alperovitz, *America Beyond Capitalism: Reclaiming Our Wealth, Our Liberty and Our Democracy* (Hoboken, NJ: Wiley, 2005); J.K.

Gibson-Graham, *A Postcapitalist Politics* (Minneapolis: University of Minnesota Press, 2007); Fred Block. *Capitalism: The Future of an Illusion* (Berkeley: University of California Press, 2018).

8 Marilyn Waring, *If Women Counted: A New Feminist Economics* (New York: HarperCollins, 1990).

9 Karl Marx, "Economic and Philosophic Manuscripts of 1844," in *The Marx-Engels Reader*, Robert Tucker, ed. (New York: Norton, 1972), 105.

10 Richard Layard, *Happiness: Lessons from a New Science* (New York: Penguin, 2005).

11 I first read this number in an unpublished piece by Gar Alperovitz. He is very trustworthy, but I found it hard to believe. So I calculated it myself and here offer how you can do so as well.

12 Karl Polanyi, *The Great Transformation: The Political and Economic Origins of Our Time* (Boston: Beacon Press, 1944).

13 Fred Block and Margaret Sommers, *The Power of Market Fundamentalism: Karl Polanyi's Critique* (Cambridge, MA: Harvard University Press, 2014). See also Kate Raworth. *Doughnut Economics: 7 Ways to Think like a 21st-Century Economist* (White River Junction, VT: Chelsea Green Publishing, 2017).

Chapter 3 Messaging That Encourages Action

1 These statements are made up and are meant to represent common climate messaging.

2 Patrick Reinsborough and Doyle Canning, *Re:Imagining Change: How to Use Story-Based Strategy to Win Campaigns, Build Movements, and Change the World*, 2nd ed. (Oakland: PM Press, 2017).

3 Jonathon P. Schuldt, Rainer Romero-Canyas, Matthew T. Ballew, and Dylan Larson-Konar, "Diverse Segments of the US Public Underestimate the Environmental Concerns of Minority and Low-Income Americans." *Proceedings of the National Academy of Sciences* 115, no. 49 (December 4, 2018): 12,429–34, first published October 29, 2018. https://doi.org/10.1073/pnas.1804698115..

4 Per Espen Stoknes, *What We Think about When We Try Not to Think about Global Warming: Toward a New Psychology of Climate Action* (White River Junction, VT: Chelsea Green Publishing, 2015), 76.

5 Stoknes, *What We Think about When We Try Not to Think about Global Warming*, 80.

6 Stoknes, 82.

7 Stoknes, 84.

Chapter 4 The Large-Scale Solutions

1 One thing to note with the number in the sections below is that these categories are overlapping. For example, the greenhouse gas emitted transporting food can count in the transportation system and also in the food system. Because I am drawing from a variety of different sources, my numbers don't work by taking the world of emissions

and cutting it into different separate pie slices. They are based on real analyses and sources are given, but they don't line up neatly. The numbers are meant to give a general sense of the scale of the sector.

2   Johan Falk, Owen Gaffney, et al., *Exponential Roadmap: Scaling 36 Solutions to Halve Emissions by 2030.* Version 1.5, 2019, https://exponentialroadmap.org/wp-content/uploads/2019/09/Exponential-Roadmap-1.5-September-19-2019.pdf.

3   David Coady, Ian Parry, Nghia-Piotr Le, and Baoping Shang, "Global Fossil Fuel Subsidies Remain Large: An Update Based on Country-Level Estimates," May 2, 2019, https://www.imf.org/en/Publications/WP/Issues/2019/05/02/Global-Fossil-Fuel-Subsidies-Remain-Large-An-Update-Based-on-Country-Level-Estimates-46509.

4   Jordan Wirfs-Brock, "Lost in Transmission: How Much Electricity Disappears Between a Power Plant and Your Plug?," Inside Energy, November 6, 2015, http://insideenergy.org/2015/11/06/lost-in-transmission-how-much-electricity-disappears-between-a-power-plant-and-your-plug/.

5   Denise Fairchild and Al Weinrub, *Energy Democracy: Advancing Equity in Clean Energy Solutions* (Washington, DC: Island Press, 2017).

6   Falk and Gaffney, *Exponential Roadmap*, 33.

7   Falk and Gaffney, 51.

8   "The Big 5—Africa's Fastest Growing Solar Energy Markets," *ESI: Africa's Power Journal*, September 26, 2019, https://www.esi-africa.com/industry-sectors/renewable-energy/the-big-5-africas-fastest-growing-solar-energy-markets-2/.

9   Falk and Gaffney, *Exponential Roadmap*, 14.

10 Falk and Gaffney, 16.

11 Falk and Gaffney, 52.

12 Falk and Gaffney, 38.

13 Paul Hawken, *Drawdown: The Most Comprehensive Plan Ever Proposed to Reverse Global Warming* (New York: Penguin, 2017), 136.

14 Hawken, *Drawdown*, 136.

15 Daniel Boffey, "Luxembourg to Become First Country to Make All Public Transport Free," *Guardian*, December 5, 2018.

16 Falk and Gaffney, *Exponential Roadmap*, 117.

17 Falk and Gaffney, 88

18 Falk and Gaffney, 90.

19 Lindsay Baker and Harvey Bernstein, "The Impact of School Buildings on Student Health and Performance," February 27, 2012, http://centerforgreenschools.org/sites/default/files/resource-files/McGrawHill_ImpactOnHealth.pdf.

20 Hawken, *Drawdown*, 103.

21 Falk and Gaffney, *Exponential Roadmap*, 120.

22 Falk and Gaffney, 120.

23 Falk and Gaffney, 80.

24 Steven J. Davis and Ken Caldeira, "Consumption-Based Accounting of CO_2 Emissions," *Proceedings of the National Academy of Sciences* 107, no. 12 (March 23, 2010): 5,687–92, https://doi.org/10.1073/pnas.0906974107.

25 Elizabeth Cline, *Overdressed: The Shockingly High Cost of Cheap Fashion* (New York: Portfolio Press, 2013).

26 Annie Leonard, *The Story of Stuff* film, 2007, https://www.storyofstuff.org/.

27 Umweltbundesamt, "Lifetime of Electrical Appliances Becoming Shorter and Shorter: Reasons for Early Replacement Are Varied—UBA Recommends a Minimum Period of Service Life," February 15, 2016, https://www.umweltbundesamt.de/en/press/pressinformation/lifetime-of-electrical-appliances-becoming-shorter.

28 Environmental Protection Agency, "Sustainable Materials Management Basics," https://www.epa.gov/smm/sustainable-materials-management-basics. Accessed December 27, 2019.

29 Falk and Gaffney, *Exponential Roadmap*, 128.

30 "Single-Use Plastic: China to Ban Bags and Other Items," BBC News, January 20, 2020, https://www.bbc.com/news/world-asia-china-51171491.

31 Falk and Gaffney, *Exponential Roadmap*, 62.

32 Falk and Gaffney, 62.

33 Falk and Gaffney, 62.

34 Intergovernmental Panel on Climate Change, *Special Report: Climate Change and Land*, chap. 5, August 2019, https://www.ipcc.ch/site/assets/uploads/2019/08/2f.-Chapter-5_FINAL.pdf.

35 Laura Stec and Eugene Cordero, *Cool Cuisine: Taking the Bite out of Global Warming* (Layton, UT: Gibbs Smith, 2008).

36 Falk and Gaffney, *Exponential Road*, 100.

37 Falk and Gaffney, 100.

38 Damian Carrington, "Healthy Diet Means a Healthy Planet, Study Shows," October 28, 2019, https://www.theguardian.com/environment/2019/oct/28/healthy-diet-means-a-healthy-planet-study-shows.

39 Sharon Omondi, "Countries Who Consume the Least Meat," WorldAtlas, August 1, 2017, https://www.worldatlas.com/articles/countries-who-consume-the-least-meat.html.

40 Jess Fanzo and Mario Herrero, "What We Eat Matters: To Change Climate Crisis, We Need to Reshape the Food System," October 8, 2019, https://www.theguardian.com/commentisfree/2019/oct/08/climate-change-food-global-heating-livestock.

41 Mary Robinson, *Climate Justice: Hope, Resilience, and the Fight for a Sustainable Future* (London: Bloomsbury, 2019), 134.

42 Falk and Gaffney, *Exponential Road*, 128.

43 Falk and Gaffney, 99.

44 Falk and Gaffney, 100.

45 Hawken, *Drawdown*, 109–10.

46 "Ethiopia 'Breaks' Tree-Planting Record to Tackle Climate Change," July 29, 2019, https://www.bbc.com/news/world-africa-49151523.

47 Sarah Koplowicz, "Using Compost for Carbon Sequestration: A Strategy for Climate Goals and Land Use Management" (master's thesis, University of San Francisco, 2019).

48 Falk and Gaffney, *Exponential Road*, 108.

49 Falk and Gaffney, 15.

50 Falk and Gaffney, 110.

51 Many people in that work use the term "development" to describe the move from poverty. It is a concept that is fraught, because development implies that everyone in the world is on one path toward an industrial lifestyle. Yet at the present, it is the most commonly used word for poverty alleviation.

52 Suzanne Goldenberg, "World Bank Rejects Energy Industry Notion That Coal Can Cure Poverty," *Guardian*, July 29, 2015, https://www.theguardian.com/environment/2015/jul/29/world-bank-coal-cure-poverty-rejects.

53 Center for Global Development, "Developed Countries Are Responsible for 79 Percent of Historical Carbon Emissions," August 8, 2015, https://www.cgdev.org/media/who-caused-climate-change-historically.

54 Falk and Gaffney, *Exponential Road*, 81–83.

55 Falk and Gaffney, 45.

56 Lighting Africa: Catalyzing Markets for Off-Grid Energy, https://www.lightingafrica.org/.

57 Falk and Gaffney, *Exponential Road*, 82.

58 CoastalChallenges, "44 Percent of Us Live in Coastal Areas," January 31, 2010, https://coastalchallenges.wordpress.com/2010/01/31/un-atlas-60-of-us-live-in-the-coastal-areas/.

59 Athlyn Cathcart-Keays, "Why Copenhagen Is Building Parks That Can Turn into Ponds," January 22, 2016, https://www.citylab.com/design/2016/01/copenhagen-parks-ponds-climate-change-community-engagement/426618/.

60 Amanda Moore, "This Wetland Restoration Project Will Help Keep the Gulf Out of New Orleans," Restore the Mississippi River Delta, July 1, 2019, http://mississippiriverdelta.org/this-wetland-restoration-project-will-help-keep-the-gulf-out-of-new-orleans/.

61 Georgetown Climate Center, "20 Good Ideas for Promoting Climate Resilience," July 2014, https://www.georgetownclimate.org/reports/20-good-ideas-for-promoting-climate-resilience.html.

62 Theresa Machemer. "How the New York City Subway Is Preparing for Climate Change," *Smithsonian Magazine*, November 26, 2019, https://www.smithsonianmag.com/smart-news/how-new-york-city-subway-preparing-climate-change-180973651/.

63 Institute for Sustainable Infrastructure, "About ISI," https://sustainableinfrastructure.org/envision/overview-of-envision/.

Chapter 5 Advice for Action

1   William Safire, "Footprint," *New York Times Magazine*, February 17, 2008, https://www.nytimes.com/2008/02/17/magazine/17wwln-safire-t.html.

2   Finis Dunaway, "The 'Crying Indian' Ad That Fooled the Environmental Movement," *Chicago Tribune*, November 21, 2017, https://www.chicagotribune.com/opinion/commentary/ct-perspec-indian-crying-environment-ads-pollution-1123-20171113-story.html.

3   Per Espen Stoknes, *What We Think about When We Try Not to Think about Global Warming: Toward a New Psychology of Climate Action* (White River Junction, VT: Chelsea Green Publishing, 2015), 91.

4   Antonio Gramsci, "Intellectuals," in *The Antonio Gramsci Reader*, David Forgacs, ed. (New York: New York University Press, 2000), 307.

5   Bill Moyer, *Doing Democracy: The MAAP Model for Organizing Social Movements* (Gabriola Island, BC, Canada: New Society Publishers, 2001).

6   Frances Fox Piven and Richard Cloward, *Poor People's Movements: Why They Succeed, How They Fail* (New York: Vintage Books, 1978).

7   Beautiful Trouble, https://beautifultrouble.org.

8   Joan Minieri and Paul Getsos, *Tools for Radical Democracy: How to Organize for Power in Your Community* (New York: Jossey-Bass, 2008).

9   Community Learning Partnership, http://communitylearningpartnership.org/.

10 Beautiful Trouble, "Toolbox," https://www.beautifultrouble.org/toolbox/.

11 Minieri and Getsos, *Tools for Radical Democracy*, chap. 7.

12 From a workshop given by Campus Camp Wellstone.

13 Marshall Ganz, "Public Narrative, Collective Action, and Power," in *Accountability through Public Opinion: From Inertia to Public Action*, Sina Odugbemi, and Taeku Lee, eds. (Washington, DC: World Bank, 2011), 273–89.

Chapter 6 Choosing the Best Policy Tools

1   Michael Hiltzik, "No Longer Termed a 'Failure,' California's Cap-and-Trade Program Faces a New Critique: Is It Too Successful?," *Los Angeles Times*, January 18, 2018, https://www.latimes.com/business/hiltzik/la-fi-hiltzik-captrade-20180111-story.html.

2   Center for Global Development, "Developed Countries Are Responsible for 79 Percent of Historical Carbon Emissions," August 8, 2015, https://www.cgdev.org/media/who-caused-climate-change-historically.

3   Atif Ansar, Ben Caldecott, and James Tilbury, "Stranded Assets and the Fossil Fuel Divestment Campaign: What Does Divestment Mean for the Valuation of Fossil Fuel Assets?," Smith School of Enterprise and the Environment, University of Oxford. Oxford, UK, October 8, 2013, https://tcdfossilfree.files.wordpress.com/2015/09/smith-school-stranded-assets.pdf.

4   Matthew Taylor and Jonathan Watts, "Revealed: The 20 Firms behind a Third of All Carbon Emissions," *Guardian*, October 9, 2019, https://www.theguardian.com/environment/2019/oct/09/revealed-20-firms-third-carbon-emissions.

5   Carlos Davidson and Cynthia Kaufman, "Is Reinvestment a Good Strategy for the Fossil Fuel Divestment Movement?," *Truthout*, January 23, 2015, https://truthout.org/articles/is-reinvestment-a-good-strategy-for-the-fossil-fuel-divestment-movement/.

Chapter 7 Finding Your People and Your Practice in the Ecosystem of the Climate Justice Movement

1   Paul Hawken, *Blessed Unrest: How the Largest Social Movement in History Is Restoring Grace, Justice, and Beauty to the World* (New York: Penguin, 2008), 142.

2   Bill Moyer, *Doing Democracy: The MAAP Model for Organizing Social Movements* (Gabriola Island, BC: New Society Publishers, 2001).

3   Moyer, *Doing Democracy*.

Chapter 8 Self-Care

1   Kai Cheng, "I Hope We Choose Love: Notes on the Application of Justice," https://medium.com/@ladysintrayda/notes-on-the-application-of-justice, May 5, 2018.

2   Adrienne Maree Brown, *Emergent Strategy* (Oakland: AK Press, 2017).

3   Wini Breines, *Community and Organization in the New Left, 1962–1968: The Great Refusal* (New York: Praeger, 1982), 6.

Annotated Bibliography

Barnes, Peter. *Climate Solutions: What Works, What Doesn't, and Why.*
An introduction to policy tools.

Beautiful Trouble. https://beautifultrouble.org.
A web-based organizing toolkit.

Brown, Adrienne Maree. *Emergent Strategy.*
A thoughtful guide to being a conscious activist.

EcoEquity. https://www.ecoequity.org.
A website focused on the global aspects of a just transition.

Engler, Paul, and Sophie Lasoff. *The Resistance Guide.*
A comprehensive online introduction to organizing for social change.

Fairchild, Denise, and Al Weinrub. *Energy Democracy: Advancing Equity in Clean Energy Solutions.*
An anthology of articles on work in the US to make the energy system more democratic and equitable.

Falk, Johan, Owen Gaffney, et al. *Exponential Roadmap: Scaling 36 Solutions to Halve Emissions by 2030.* https://exponentialroadmap.org/wp-content/uploads/2019/09Exponential-Roadmap-1.5-September-19-2019.pdf.
A deep dive into the path to rapid decreases in emissions.

Hawken, Paul, ed. *Drawdown: The Most Comprehensive Plan Ever Proposed to Reverse Global Warming.*
This is a beautiful and very well researched book that analyzes the top one hundred most impactful changes that are needed to get to net zero carbon emissions.

Kaufman, Cynthia. *Getting Past Capitalism: History, Vision, Hope.*
An analysis of how to challenge capitalism one step at a time.

Klein, Naomi. *This Changes Everything: Capitalism vs. the Climate.*
An introduction to the intersections of capitalism and climate justice work and a great read.

LaDuke, Winona. *The Winona LaDuke Chronicles: Stories from the Front Lines in the Battle for Environmental Justice.*
A collection of essays by LaDuke on indigenous climate action.

Latour, Bruno. *Down to Earth: Politics in the New Climatic Regime.*
An analysis of the present political moment which links the climate crisis, the rise of ethnonationalism, and the global refugee crisis.

Leonard, Annie. *The Story of Stuff.*
A twenty-minute animated film on the wasteful nature of a consumer economy. It is fun and accessible and has a lot of analysis behind it. You can find it and a series of subsequent films and articles at: The Story of Stuff Project.

McKibben, Bill, ed. *The Global Warming Reader: A Century of Writing about Climate Change.*
A comprehensive introduction to the subject with many important speeches and documents.

———. "Global Warming's Terrifying Math."
This article kicked off the fossil fuel divestment movement. It is clear, concise, and very compelling.

Minieri, Joan, and Paul Getsos. *Tools for Radical Democracy: How to Organize for Power in Your Community.*
A comprehensive toolkit for community organizing.

Raworth, Kate. *Doughnut Economics: 7 Ways to Think like a 21st-Century Economist.*
An introduction to thinking about economics as a practice of building social and environmental sustainability.

Reinsborough, Patrick, and Doyle Canning. *Re:Imagining Change: How to Use Story-Based Strategy to Win Campaigns, Build Movements, and Change the World.*
A guide for organizing that is grounded in story-based strategy.

Romm, Joseph. *Climate Change: What Everyone Needs to Know.*
An encyclopedia of climate facts. It is well indexed and is organized as answers to frequently asked questions.

Bibliography

Alperovitz, Gar. *America Beyond Capitalism: Reclaiming Our Wealth, Our Liberty, and Our Democracy.* Hoboken, NJ: Wiley, 2005.

Ansar, Atif, Ben Caldecott, and James Tilbury. "Stranded Assets and the Fossil Fuel Divestment Campaign: What Does Divestment Mean for the Valuation of Fossil Fuel Assets?" Smith School of Enterprise and the Environment, University of Oxford. October 8, 2013. https://www.smithschool.ox.ac.uk/publications/reports/SAP-divestment-report-final.pdf.

ARO Watch. "Tracking the Trillion Dollar Retirement of Oil." https://www.arowatch.org/.

Baker, Lindsay, and Harvey Bernstein. "The Impact of School Buildings on Student Health and Performance." February 27, 2012. http://centerforgreenschools.org/sites/default/files/resource-files/McGrawHill_ImpactOnHealth.pdf.

BBC. "Ethiopia 'Breaks' Tree-Planting Record to Tackle Climate Change." July 29, 2019. https://www.bbc.com/news/world-africa-49151523.

Blackburn, Robin. *The Making of New World Slavery: From the Baroque to the Modern, 1492–1800.* New York: Verso, 1997.

Block, Fred. *Capitalism: The Future of an Illusion.* Berkeley: University of California Press, 2018.

Block, Fred, and Margaret Sommers. *The Power of Market Fundamentalism: Karl Polanyi's Critique.* Cambridge, MA: Harvard University Press, 2014.

Boffey, Daniel. "Luxembourg to Become First Country to Make All Public Transport Free." *Guardian,* December 5, 2018.

Breines, Wini. *Community and Organization in the New Left, 1962–1968: The Great Refusal.* New York: Praeger, 1982.

Carrington, Damian. "Healthy Diet Means a Healthy Planet, Study Shows." *Guardian,* October 28, 2019. https://www.theguardian.com/environment/2019/oct/28/healthy-diet-means-a-healthy-planet-study-shows.

Center for Global Development. "Developed Countries Are Responsible for 79 Percent of Historical Carbon Emissions." August 8, 2015. https://www.cgdev.org/media/who-caused-climate-change-historically.

Cheng, Kai, "I Hope We Choose Love: Notes on the Application of Justice." May 5, 2018. https://medium.com/@ladysintrayda/notes-on-the-application-of-justice.

Clayton, Susan, et al. *Mental Health and Our Changing Climate: Impacts, Implications, and Guidance.* American Psychological Association, March 2017. https://www.apa.org/news/press/releases/2017/03/mental-health-climate.pdf.

Climate Justice Alliance. "Just Transition." https://climatejusticealliance.org/just-transition/.

Cline, Elizabeth. *Overdressed: The Shockingly High Cost of Cheap Fashion.* New York: Portfolio Press, 2013.

CoastalChallenges. "44 Percent of Us Live in Coastal Areas." January 31, 2010. https://coastalchallenges.wordpress.com/2010/01/31/un-atlas-60-of-us-live-in-the-coastal-areas/.

Community Learning Partnership. http://communitylearningpartnership.org/.

Davidson, Carlos, and Cynthia Kaufman. "Is Reinvestment a Good Strategy for the Fossil Fuel Divestment Movement?" January 23, 2015. *Truthout.* https://truthout.org/articles/is-reinvestment-a-good-strategy-for-the-fossil-fuel-divestment-movement/.

Davis, Steven J., and Ken Caldeira. "Consumption-Based Accounting of CO_2 Emissions." *Proceedings of the National Academy of Sciences* 107, no. 12 (March 23, 2010): 5,687–92. https://doi.org/10.1073/pnas.0906974107.

Dunaway, Finis. "The 'Crying Indian' Ad That Fooled the Environmental Movement." *Chicago Tribune.* November 21, 2017. https://www.chicagotribune.com/opinion/commentary/ct-perspec-indian-crying-environment-ads-pollution-1123-20171113-story.html.

Di Christopher, Tom. "Climate Disasters Cost the World $650 Billion over 3 Years—Americans Are Bearing the Brunt: Morgan Stanley." February 14, 2019: https://www.cnbc.com/2019/02/14/climate-disasters-cost-650-billion-over-3-years-morgan-stanley.html.

Dickinson, Tim. "Study: U.S. Fossil Fuel Subsidies Exceed Pentagon Spending." *Rolling Stone*, May 8, 2019. https://www.rollingstone.com/politics/politics-news/fossil-fuel-subsidies-pentagon-spending-imf-report-833035/.

Durkin Erin. "North Carolina Didn't Like Science on Sea Levels . . . So Passed a Law against It." *Guardian*, September 12, 2018. https://www.theguardian.com/us-news/2018/sep/12/north-carolina-didnt-like-science-on-sea-levels-so-passed-a-law-against-it.

Environmental Protection Agency. "Sustainable Materials Management Basics." https://www.epa.gov/smm/sustainable-materials-management-basics.

Fairchild, Denise, and Al Weinrub. *Energy Democracy: Advancing Equity in Clean Energy Solutions*. Washington, DC: Island Press, 2017.

Falk, Johan, Owen Gaffney, et al. *Exponential Roadmap: Scaling 36 Solutions to Halve Emissions by 2030*. Version 1.5, 2019. https://exponentialroadmap.org/wp-content/uploads/2019/09/Exponential-Roadmap-1.5-September-19-2019.pdf.

Fanzo, Jess, and Mario Herrero. "What We Eat Matters: To Change Climate Crisis, We Need to Reshape the Food System." *Guardian*, October 8, 2019. https://www.theguardian.com/commentisfree/2019/oct/08/climate-change-food-global-heating-livestock?

Ganz, Marshall. "Public Narrative, Collective Action, and Power." In *Accountability through Public Opinion: From Inertia to Public* Action, edited by Sina Odugbemi and Taeku Lee, 273–89. Washington, DC: World Bank, 2011.

Georgetown Climate Center. "20 Good Ideas for Promoting Climate Resilience." https://www.georgetownclimate.org/files/report/GCC-20%20Good%20Ideas-July%202014.pdf.

Gibson-Graham, J.K.. *A Postcapitalist Politics*. Minneapolis: University of Minnesota Press, 2007.

Goldenberg, Suzanne. "World Bank Rejects Energy Industry Notion That Coal Can Cure Poverty." *Guardian*, July 29, 2015. https://www.theguardian.com/environment/2015/jul/29/world-bank-coal-cure-poverty-rejects.

Gramsci, Antonio. "Intellectuals." In *The Antonio Gramsci Reader*, edited by David Forgacs. New York: New York University Press, (1935) 2000.

Harvey, David. *A Brief History of Neoliberalism*. Oxford: Oxford University Press. 2007.

Hawken, Paul. *Blessed Unrest: How the Largest Social Movement in History Is Restoring Grace, Justice, and Beauty to the World*. New York: Penguin, 2008.

———. *Drawdown: The Most Comprehensive Plan Ever Proposed to Reverse Global Warming*. New York: Penguin, 2017.

Hiltzik, Michael. "No Longer Termed a 'Failure,' California's Cap-and-Trade Program Faces a New Critique: Is It Too Successful?" *Los Angeles Times*, January 18, 2018. https://www.latimes.com/business/hiltzik/la-fi-hiltzik-captrade-20180111-story.html.

Institute for Sustainable Infrastructure. "About Envision." https://sustainableinfrastructure.org/envision/overview-of-envision/.

Intergovernmental Panel on Climate Change. *Special Report: Climate Change and Land*. Chapter 5. August 2019. https://www.ipcc.ch/site/assets/uploads/2019/08/2f.-Chapter-5_FINAL.pdf.

International Institute for Sustainable Development, "Reforming Subsidies Could Pay for a Clean Energy Revolution," updated August 8, 2019, https://www.iisd.org/gsi/news-events/reforming-subsidies-could-help-pay-clean-energy-revolution-report.

Katznelson, Ira. *When Affirmative Action Was White: An Untold History of Racial Inequality in Twentieth-Century America*. New York: W.W. Norton, 2006.

Kaufman, Cynthia. *Getting Past Capitalism: History, Vision, Hope*. Lanham, MD: Lexington Books, 2012.

Klein, Naomi. *This Changes Everything: Capitalism vs the Climate*. New York: Simon and Schuster, 2015.

Koplowicz, Sarah. "Using Compost for Carbon Sequestration: A Strategy for Climate Goals and Land Use Management." Master's thesis, University of San Francisco, 2019.

Labor Network for Sustainability. "'Just Transition'—Just What Is It?" https://www.labor4sustainability.org/uncategorized/just-transition-just-what-is-it/. Accessed May 2, 2019.

Latour, Bruno. *Down to Earth: Politics in the New Climatic Regime*. Cambridge: Wiley, 2018.

Layard, Richard. *Happiness: Lessons from a New Science*. New York: Penguin, 2005.

Leonard, Annie. *The Story of Stuff* film. 2007. https://www.storyofstuff.org/.

Lighting Africa: Catalyzing Markets for Off-Grid Energy. https://www.lightingafrica.org/.

Machemer, Theresa. "How the New York City Subway Is Preparing for Climate Change." *Smithsonian Magazine*, November 26, 2019. https://www.smithsonianmag.com/smart-news/how-new-york-city-subway-preparing-climate-change-180973651/.

Marx, Karl. "Economic and Philosophic Manuscripts of 1844." In *The Marx-Engels Reader*, edited by Robert Tucker, 66–125. New York: Norton, 1972.

Mayer, Jane. *Dark Money: The Hidden History of the Billionaires behind the Rise of the Radical Right*. New York: Doubleday, 2016.

Minieri, Joan, and Paul Getsos. *Tools for Radical Democracy: How to Organize for Power in Your Community*. New York: Jossey-Bass, 2008.

Monbiot, George. *Heat: How to Stop the Planet Burning*. New York: Penguin, 2007.

Moore, Amanda. "This Wetland Restoration Project Will Help Keep the Gulf out of New Orleans." Restore the Mississippi River Delta, July 1, 2019. http://mississippiriverdelta.org/this-wetland-restoration-project-will-help-keep-the-gulf-out-of-new-orleans/.

Moyer, Bill. *Doing Democracy: The MAAP Model for Organizing Social Movements*. Gabriola Island, BC, Canada: New Society Publishers, 2001.

National Network for Immigrant and Refugee Rights. "Fact Sheet: Climate Change, Global Migration, and Human Rights." September 2018. https://www.nnirr.org/~nnirrorg/drupal/sites/default/files/climate_change_and_migration_fact_sheet_final.pdf.

Oil Change International. "Fossil Fuel Subsidies Overview." http://priceofoil.org/fossil-fuel-subsidies/. Accessed December 12, 2019.

Omondi, Sharon. "Countries Who Consume the Least Meat." WorldAtlas, August 1, 2017, https://www.worldatlas.com/articles/countries-who-consume-the-least-meat.html.

Piven, Frances Fox, and Richard Cloward. *Poor People's Movements: Why They Succeed, How They Fail.* New York: Vintage Books, 1978.

Polanyi, Karl. *The Great Transformation: The Political and Economic Origins of Our Time.* Boston: Beacon Press, 1944.

Raworth, Kate. *Doughnut Economics: 7 Ways to Think like a 21st-Century Economist.* White River Junction Vermont: Chelsea Green Publishing, 2017.

Reinsborough, Patrick, and Doyle Canning. *Re:Imagining Change: How to Use Story-Based Strategy to Win Campaigns, Build Movements, and Change the World.* 2nd ed. Oakland: PM Press, 2017.

Robinson, Mary. *Climate Justice: Hope, Resilience, and the Fight for a Sustainable Future.* London: Bloomsbury, 2019.

Rojas, Maisa. "The Climate Crisis and Inequality a Recipe for Disaster: Just Look at Chile." *Guardian Weekly*, December 13, 2019. https://www.theguardian.com/commentisfree/2019/dec/08/un-conference-global-heating-cop25-chile-madrid-climate-crisis.

Romm, Joseph. *Climate Change: What Everyone Needs to Know.* 2nd ed. Oxford: Oxford University Press, 2018.

Safire, William. "Footprint." *New York Times Magazine*, February 17, 2008. https://www.nytimes.com/2008/02/17/magazine/17wwln-safire-t.html.

Schuldt, Jonathon P., Rainer Romero-Canyas, Matthew T. Ballew, and Dylan Larson-Konar. "Diverse Segments of the US Public Underestimate the Environmental Concerns of Minority and Low-Income Americans." *Proceedings of the National Academy of Sciences* 115, no. 49 (December 4, 2018): 12,429–34, first published October 29, 2018. https://doi.org/10.1073/pnas.1804698115.

"Single-Use Plastic: China to Ban Bags and Other Items." January 20, 2020. BBC News. https://www.bbc.com/news/world-asia-china-51171491.

Stec, Laura, and Eugene Cordero. *Cool Cuisine: Taking the Bite out of Global Warming.* Layton, UT: Gibbs Smith, 2008.

Stern, Nicholas. *The Economics of Climate Change: The Stern Review*, Cambridge: Cambridge University Press, 2007.

Stoknes, Per Espen. *What We Think about When We Try Not to Think about Global Warming: Toward a New Psychology of Climate Action.* White River Junction, VT: Chelsea Green Publishing, 2015.

Taylor, Matthew, and Jonathan Watts. "Revealed: The 20 Firms behind a Third of All Carbon Emissions." *Guardian*, October 9, 2019. https://www.theguardian.com/environment/2019/oct/09/revealed-20-firms-third-carbon-emissions.

Umweltbundesamt. "Lifetime of Electrical Appliances Becoming Shorter and Shorter: Reasons for Early Replacement Are Varied—UBA Recommends a Minimum Period of Service Life." February 15, 2016. https://www.umweltbundesamt.de/en/press/pressinformation/lifetime-of-electrical-appliances-becoming-shorter.

Union of Concerned Scientists. "2.5 Million Homes, Businesses Totaling $1 Trillion Threatened by High Tide Flooding." June 18, 2018. https://www.ucsusa.org/about/news/25-million-homes-threatened-high-tide-flooding.

Waring, Marilyn. *If Women Counted: A New Feminist Economics*. New York: HarperCollins, 1990.

Wirfs-Brock, Jordan. "Lost in Transmission: How Much Electricity Disappears Between a Power Plant and Your Plug?," November 6, 2015. Inside Energy. http://insideenergy.org/2015/11/06/lost-in-transmission-how-much-electricity-disappears-between-a-power-plant-and-your-plug/. Accessed December 27, 2019.

About the Authors

Cynthia Kaufman is the director of the Vasconcellos Institute for Democracy in Action at De Anza College, where she runs and teaches in a community organizer training program. She is the author of three previous books on social change: *Challenging Power: Democracy and Accountability in a Fractured World* (Bloomsbury, 2020); *Ideas for Action: Relevant Theory for Radical Change* (2nd ed. PM Press, 2016); and *Getting Past Capitalism: History, Vision, Hope* (Lexington Books, 2012). She has been active in a wide variety of social justice movements including Central American solidarity, union organizing, police accountability, and most recently tenants' rights and climate change. She publishes on social justice at *Common Dreams*.

Bill McKibben is an author and environmentalist who in 2014 was awarded the Right Livelihood Prize, sometimes called the "alternative Nobel." His 1989 book *The End of Nature* is regarded as the first book for a general audience about climate change and has appeared in twenty-four languages. He's gone on to write a dozen more books. He is a founder of 350.org, the first planet-wide, grassroots climate change movement, which has organized twenty thousand rallies around the world in every country save North Korea, spearheaded the resistance to the Keystone Pipeline, and launched the fast-growing fossil fuel divestment movement. He is the author of *Falter, Oil and Honey, Eaarth, Deep Economy, The End of Nature*, and numerous others.

ABOUT PM PRESS

PM Press is an independent, radical publisher of books and media to educate, entertain, and inspire. Founded in 2007 by a small group of people with decades of publishing, media, and organizing experience, PM Press amplifies the voices of radical authors, artists, and activists. Our aim is to deliver bold political ideas and vital stories to all walks of life and arm the dreamers to demand the impossible. We have sold millions of copies of our books, most often one at a time, face to face. We're old enough to know what we're doing and young enough to know what's at stake. Join us to create a better world.

PM Press
PO Box 23912
Oakland, CA 94623
www.pmpress.org

PM Press in Europe
europe@pmpress.org
www.pmpress.org.uk

FRIENDS OF PM PRESS

These are indisputably momentous times—the financial system is melting down globally and the Empire is stumbling. Now more than ever there is a vital need for radical ideas.

In the years since its founding—and on a mere shoestring—PM Press has risen to the formidable challenge of publishing and distributing knowledge and entertainment for the struggles ahead. With over 450 releases to date, we have published an impressive and stimulating array of literature, art, music, politics, and culture. Using every available medium, we've succeeded in connecting those hungry for ideas and information to those putting them into practice.

Friends of PM allows you to directly help impact, amplify, and revitalize the discourse and actions of radical writers, filmmakers, and artists. It provides us with a stable foundation from which we can build upon our early successes and provides a much-needed subsidy for the materials that can't necessarily pay their own way. You can help make that happen—and receive every new title automatically delivered to your door once a month—by joining as a Friend of PM Press. And, we'll throw in a free T-shirt when you sign up.

Here are your options:

- **$30 a month** Get all books and pamphlets plus 50% discount on all webstore purchases

- **$40 a month** Get all PM Press releases (including CDs and DVDs) plus 50% discount on all webstore purchases

- **$100 a month** Superstar—Everything plus PM merchandise, free downloads, and 50% discount on all webstore purchases

For those who can't afford $30 or more a month, we have **Sustainer Rates** at $15, $10 and $5. Sustainers get a free PM Press T-shirt and a 50% discount on all purchases from our website.

Your Visa or Mastercard will be billed once a month, until you tell us to stop. Or until our efforts succeed in bringing the revolution around. Or the financial meltdown of Capital makes plastic redundant. Whichever comes first.

Ideas for Action: Relevant Theory for Radical Change, 2nd Ed.

Cynthia Kaufman

ISBN: 978-1-62963-147-9
$22.95   352 pages

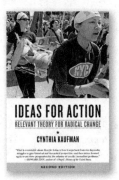

Written in an engaging and accessible style, *Ideas for Action* gives activists the intellectual tools to turn discontent into a plan of action.
Exploring a wide range of political traditions—including Marxism, anarchism, anti-imperialism, postmodernism, feminism, critical race theory, and environmentalism—Cynthia Kaufman acknowledges the strengths and weaknesses of a variety of political movements and the ideologies inspired by or generated through them. Kaufman incorporates elements of her own activist experiences and presents a coherent analysis without pretending to offer "the final word" on complex issues. Instead, she helps orient a critical understanding of the social world and a glimpse of the excitement and rewards of serious intellectual engagement with political ideas.

Fully updated to confront pressing issues of today—from mass incarceration to climate change, from the war on terror to the national security state, from rising inequality to a global shortage of care, Ideas for Action also examines the work of diverse thinkers such as Adam Smith, Paulo Freire, Grace Lee Boggs, and Stuart Hall. Kaufman's insights break the chains of cynicism and lay a foundation for more effective organizing.

"There's a long-running conversation about what's wrong with our world and how to fix it. Ideas for Action *fills in the 'backstory' that can help you to join that conversation."*
—Jeremy Brecher, author of *Strike!*

"What is remarkable about Cynthia Kaufman's book Ideas for Action *is how it steps back from our day-to-day struggles to gain historical and theoretical perspective, and then moves forward again to use these perspectives for the solution of specific, immediate problems. The book ranges broadly over many contemporary problems, and manages to be both theoretical and practical in the analysis of these problems."*
—Howard Zinn, author of *A People's History of the United States*

Against Doom: A Climate Insurgency Manual

Jeremy Brecher

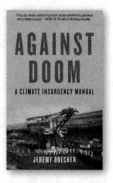

ISBN: 978-1-62963-385-5
$12.95   128 pages

Before the election of Donald Trump the world was already speeding toward climate catastrophe. Now President Trump has jammed his foot on the global warming accelerator. Is there any way for the rest of us to put on the brakes?

Climate insurgency is a strategy for using people power to realize our common interest in protecting the climate. It uses mass, global, nonviolent action to challenge the legitimacy of public and corporate officials who are perpetrating climate destruction.

A global climate insurgency has already begun. It has the potential to halt and roll back Trump's fossil fuel agenda and the global thrust toward climate destruction.

Against Doom: A Climate Insurgency Manual tells how to put that strategy into action—and how it can succeed. It is a handbook for halting global warming and restoring our climate—a how-to for climate insurgents.

"*Against Doom lays out key elements of a far-reaching, global-scaled, pragmatic, people-powered strategy to topple the power of the fossil fuel industry and the institutions behind it.*"
—David Solnit, author of *Globalize Liberation: How to Uproot the System and Build a Better World*

"*In* Against Doom, *Brecher has provided the climate movement with two essential tools: a moral framework for the struggle against fossil fuels, and an actual plan for victory. By blending sober social movement analysis with the fire of grassroots activism, this book shows that there is a genuine, and winnable, case against the fossil fuel economy—a case to be argued in the streets as well as the courtroom. It's an essential volume for anyone committed to social change in the fight against climate change.*"
—Joseph Hamilton, Climate Defense Project

Save the Humans?
Common Preservation in Action

Jeremy Brecher

ISBN: 978-1-62963-798-3
$20.00   272 pages

We the people of the world are creating the
conditions for our own self-extermination,
whether through the bang of a nuclear
holocaust or the whimper of an expiring ecosphere. Today our individual
self-preservation depends on common preservation—cooperation to
provide for our mutual survival and well-being.

For half a century Jeremy Brecher has been studying and participating in
social movements that have created new forms of common preservation.
Through entertaining storytelling and personal narrative, *Save the
Humans?* provides a unique and revealing interpretation of how social
movements arise and how they change the world. Brecher traces a path
that leads from the sitdown strikes on the pyramids of ancient Egypt
through America's mass strikes and labor revolts to the struggle against
economic globalization to today's battles against climate change.

Weaving together personal experience, scholarly research, and historical
interpretation, Jeremy Brecher shows how we can construct a "human
survival movement" that could "save the humans." He sums up the
theme of this book: "I have seen common preservation—and it works."
For those seeking an understanding of social movements and an
alternative to denial and despair, there is simply no better place to look
than *Save the Humans?*

*"This is a remarkable book: part personal story, part intellectual history
told in the first person by a skilled writer and assiduous historian, part
passionate but clearly and logically argued plea for pushing the potential
of collective action to preserve the human race. Easy reading and full of
useful and unforgettable stories. . . . A medicine against apathy and political
despair much needed in the U.S. and the world today."*
—Peter Marcuse, author of *Cities for People, Not for Profit: Critical Urban
Theory*

Common Preservation: In a Time of Mutual Destruction

Jeremy Brecher
with a Foreword by Todd Vachon

ISBN: 978-1-62963-788-4
$26.95   400 pages

As world leaders eschew cooperation to
address climate change, nuclear proliferation,
economic meltdown, and other threats to
our survival, more and more people experience a pervasive sense of
dread and despair. Is there anything we can do? What can put us on the
course from mutual destruction to common preservation? In the past,
social movements have sometimes made rapid and unexpected changes
that countered apparently incurable social problems. Jeremy Brecher
presents scores of historical examples of people who changed history
by adopting strategies of common preservation, showing what we can
we learn from past social movements to better confront today's global
threats of climate change, war, and economic chaos.

In *Common Preservation*, Brecher shares his experiences and what he has
learned that can help ward off mutual destruction and provides a unique
heuristic—a tool kit for thinkers and activists—to understand and create
new forms of common preservation.

"Jeremy Brecher's work is astonishing and refreshing; and, God knows,
necessary."
—Studs Terkel

"Chapter by chapter, I learn from it; and I admire its ambition. When I
sampled it, it engaged me so much that I set aside other work until I finished
it. Overall, a fine manuscript. Rich in content. Also engaging. Is it not all or
part of a philosophy or worldview?"
—Charles Lindblom, Sterling Professor Emeritus of Political Science and
Economics at Yale University; author of *The Market System*

"It is an autobiography of intellectual exploration and of practical
experimentation with the problems of social injustice. It is a project of the
urgent transmission of the lessons learned undertaken under the duress
of historical time which threatens catastrophe. It is a valedictory and an
exhortation."
—Joshua Dubler, Society of Fellows, Columbia University; author of
Down in the Chapel: Religious Life in an American Prison

A Line in the Tar Sands: Struggles for Environmental Justice

Edited by Joshua Kahn, Stephen D'Arcy, Tony Weis, Toban Black
with a Foreword by Naomi Klein
and Bill McKibben

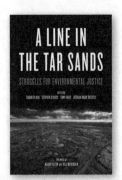

ISBN: 978-1-62963-039-7
$24.95   392 pages

Tar sands "development" comes with an enormous environmental
and human cost. In the tar sands of Alberta, the oil industry is using
vast quantities of water and natural gas to produce synthetic crude
oil, creating drastically high levels of greenhouse gas emissions
and air and water pollution. But tar sands opponents—fighting a
powerful international industry—are likened to terrorists, government
environmental scientists are muzzled, and public hearings are concealed
and rushed.

Yet, despite the formidable political and economic power behind the
tar sands, many opponents are actively building international networks
of resistance, challenging pipeline plans while resisting threats to
Indigenous sovereignty and democratic participation. Including leading
voices involved in the struggle against the tar sands, *A Line in the
Tar Sands* offers a critical analysis of the impact of the tar sands and
the challenges opponents face in their efforts to organize effective
resistance.

Contributors include: Greg Albo, Sâkihitowin Awâsis, Toban Black, Rae
Breaux, Jeremy Brecher, Linda Capato, Jesse Cardinal, Angela V. Carter,
Emily Coats, Stephen D'Arcy, Yves Engler, Cherri Foytlin, Sonia Grant,
Harjap Grewal, Randolph Haluza-DeLay, Ryan Katz-Rosene, Naomi
Klein, Melina Laboucan-Massimo, Winona LaDuke, Crystal Lameman,
Christine Leclerc, Kerry Lemon, Matt Leonard, Martin Lukacs, Tyler
McCreary, Bill McKibben, Yudith Nieto, Joshua Kahn Russell, Macdonald
Stainsby, Clayton Thomas-Muller, Brian Tokar, Dave Vasey, Harsha
Walia, Tony Weis, Rex Weyler, Will Wooten, Jess Worth, and Lilian Yap.

*The editors' proceeds from this book will be donated to frontline grassroots
environmental justice groups and campaigns.*

Anthropocene or Capitalocene? Nature, History, and the Crisis of Capitalism

Edited by Jason W. Moore

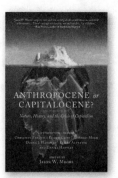

ISBN: 978-1-62963-148-6
$21.95   304 pages

The Earth has reached a tipping point.
Runaway climate change, the sixth great extinction of planetary life, the acidification of the oceans—all point toward an era of unprecedented turbulence in humanity's relationship within the web of life. But just what is that relationship, and how do we make sense of this extraordinary transition?

Anthropocene or Capitalocene? offers answers to these questions from a dynamic group of leading critical scholars. They challenge the theory and history offered by the most significant environmental concept of our times: the Anthropocene. But are we living in the Anthropocene, literally the "Age of Man"? Is a different response more compelling, and better suited to the strange—and often terrifying—times in which we live? The contributors to this book diagnose the problems of Anthropocene thinking and propose an alternative: the global crises of the twenty-first century are rooted in the Capitalocene; not the Age of Man but the Age of Capital.

Anthropocene or Capitalocene? offers a series of provocative essays on nature and power, humanity, and capitalism. Including both well-established voices and younger scholars, the book challenges the conventional practice of dividing historical change and contemporary reality into "Nature" and "Society," demonstrating the possibilities offered by a more nuanced and connective view of human environment-making, joined at every step with and within the biosphere. In distinct registers, the authors frame their discussions within a politics of hope that signal the possibilities for transcending capitalism, broadly understood as a "world-ecology" that joins nature, capital, and power as a historically evolving whole.

Contributors include Jason W. Moore, Eileen Crist, Donna J. Haraway, Justin McBrien, Elmar Altvater, Daniel Hartley, and Christian Parenti.

Re:Imagining Change: How to Use Story-Based Strategy to Win Campaigns, Build Movements, and Change the World

Patrick Reinsborough &
Doyle Canning

ISBN: 978-1-62963-384-8
$18.95   224 pages

Re:Imagining Change provides resources, theory, hands-on tools, and illuminating case studies for the next generation of innovative change-makers. This unique book explores how culture, media, memes, and narrative intertwine with social change strategies, and offers practical methods to amplify progressive causes in the popular culture.

Re:Imagining Change is an inspirational inside look at the trailblazing methodology developed by the Center for Story-based Strategy over fifteen years of their movement building partnerships. This practitioner's guide is an impassioned call to innovate our strategies for confronting the escalating social and ecological crises of the twenty-first century. This new, expanded second edition includes updated examples from the frontlines of social movements and provides the reader with easy-to-use tools to change the stories they care about most.

"All around us the old stories are failing, crumbling in the face of lived experience and scientific reality. But what stories will replace them? That is the subject of this crucial book: helping readers to tell irresistible stories about deep change—why it is needed and what it will look like. The Story-based Strategy team has been doing this critical work for fifteen years, training an entire generation in transformative communication. This updated edition of Re:Imagining Change *is a thrilling addition to the activist tool kit."*
—Naomi Klein, author of *This Changes Everything: Capitalism vs. the Climate*

"This powerful and useful book is an invitation to harness the transformative power of stories by examining social change strategy through the lens of narrative. Re:Imagining Change *is an essential resource to make efforts for fundamental social change more enticing, compelling, and effective. It's a potent how-to book for anyone working to create a better world."*
—Ilyse Hogue, president, NARAL Pro-Choice America

Facebooking the Anthropocene in Raja Ampat: Technics and Civilization in the 21st Century

Bob Ostertag

ISBN: 978-1-62963-830-0
$17.00   192 pages

The three essays of *Facebooking the Anthropocene in Raja Ampat* paint a deeply intimate portrait of the cataclysmic shifts between humans, technology, and the so-called natural world. Amid the breakneck pace of both technological advance and environmental collapse, Bob Ostertag explores how we are changing as fast as the world around us—from how we make music, to how we have sex, to what we do to survive, and who we imagine ourselves to be. And though the environmental crisis terrifies and technology overwhelms, Ostertag finds enough creativity, compassion, and humor in our evolving behavior to keep us laughing and inspired as the world we are building overtakes the world we found.

A true polymath who covered the wars in Central America during the 1980s, recorded dozens of music projects, and published books on startlingly eclectic subjects, Ostertag fuses his travels as a touring musician with his journalist's eye for detail and the long view of a historian. Wander both the physical and the intellectual world with him. Watch Buddhist monks take selfies while meditating and DJs who make millions of dollars pretend to turn knobs in front of crowds of thousands. Shiver with families huddling through the stinging Detroit winter without heat or electricity. Meet Spice Islanders who have never seen flushing toilets yet have gay hookup apps on their phones.

Our best writers have struggled with how to address the catastrophes of our time without looking away. Ostertag succeeds where others have failed, with the moral acuity of Susan Sontag, the technological savvy of Lewis Mumford, and the biting humor of Jonathan Swift.

"With deep intelligence and an acute and off-center sensibility, Robert Ostertag gives us a riveting and highly personalized view of globalization, from the soaring skyscapes of Shanghai to the darkened alleys of Yogyakarta."
—Frances Fox Piven, coauthor of *Regulating the Poor* and *Poor People's Movements*

Catastrophism: The Apocalyptic Politics of Collapse and Rebirth

Sasha Lilley, David McNally, Eddie Yuen, and James Davis with a foreword by Doug Henwood

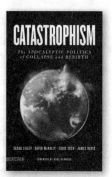

ISBN: 978-1-60486-589-9
$16.00   192 pages

We live in catastrophic times. The world is reeling from the deepest economic crisis since the Great Depression, with the threat of further meltdowns ever-looming. Global warming and myriad dire ecological disasters worsen—with little if any action to halt them—their effects rippling across the planet in the shape of almost biblical floods, fires, droughts, and hurricanes. Governments warn that no alternative exists than to take the bitter medicine they prescribe—or risk devastating financial or social collapse. The right, whether religious or secular, views the present as catastrophic and wants to turn the clock back. The left fears for the worst, but hopes some good will emerge from the rubble. Visions of the apocalypse and predictions of impending doom abound. Across the political spectrum, a culture of fear reigns.

Catastrophism explores the politics of apocalypse—on the left and right, in the environmental movement, and from capital and the state—and examines why the lens of catastrophe can distort our understanding of the dynamics at the heart of these numerous disasters—and fatally impede our ability to transform the world. Lilley, McNally, Yuen, and Davis probe the reasons why catastrophic thinking is so prevalent, and challenge the belief that it is only out of the ashes that a better society may be born. The authors argue that those who care about social justice and the environment should eschew the Pandora's box of fear—even as it relates to indisputably apocalyptic climate change. Far from calling people to arms, they suggest, catastrophic fear often results in passivity and paralysis—and, at worst, reactionary politics.

"This groundbreaking book examines a deep current—on both the left and right—of apocalyptical thought and action. The authors explore the origins, uses, and consequences of the idea that collapse might usher in a better world. Catastrophism *is a crucial guide to understanding our tumultuous times, while steering us away from the pitfalls of the past."*
—Barbara Epstein, author of *Political Protest and Cultural Revolution: Nonviolent Direct Action in the 1970s and 1980s*